SpringerBriefs in Applied Sciences and Technology

Thermal Engineering and Applied Science

Series editor

Francis A. Kulacki, University of Minnesota, Minneapolis, MN, USA

SpringerBriefs present concise summaries of cutting-edge research and practical applications across a wide spectrum of fields. Featuring compact volumes of 50–125 pages, the series covers a range of content from professional to academic.

Typical publications can be:

- A timely report of state-of-the art methods
- An introduction to or a manual for the application of mathematical or computer techniques
- A bridge between new research results, as published in journal articles
- A snapshot of a hot or emerging topic
- An in-depth case study
- A presentation of core concepts that students must understand in order to make independent contributions

SpringerBriefs are characterized by fast, global electronic dissemination, standard publishing contracts, standardized manuscript preparation and formatting guidelines, and expedited production schedules.

On the one hand, **SpringerBriefs in Applied Sciences and Technology** are devoted to the publication of fundamentals and applications within the different classical engineering disciplines as well as in interdisciplinary fields that recently emerged between these areas. On the other hand, as the boundary separating fundamental research and applied technology is more and more dissolving, this series is particularly open to trans-disciplinary topics between fundamental science and engineering.

Indexed by EI-Compendex, SCOPUS and Springerlink.

More information about this series at http://www.springer.com/series/8884

Hitoshi Sakamoto • Francis A. Kulacki

Buoyancy-Driven Flow in Fluid-Saturated Porous Media near a Bounding Surface

 Springer

Hitoshi Sakamoto
Japan Research Center
Huawei Technologies Japan
Yokohama, Japan

Francis A. Kulacki
Department of Mechanical Engineering
University of Minnesota
Minneapolis, MN, USA

ISSN 2191-530X   ISSN 2191-5318   (electronic)
SpringerBriefs in Applied Sciences and Technology
ISSN 2193-2530   ISSN 2193-2549   (electronic)
SpringerBriefs in Thermal Engineering and Applied Science
ISBN 978-3-319-89886-5   ISBN 978-3-319-89887-2   (eBook)
https://doi.org/10.1007/978-3-319-89887-2

Library of Congress Control Number: 2018938804

Printed on acid-free paper

This Springer imprint is published by the registered company Springer International Publishing AG part of Springer Nature.
The registered company address is: Gewerbestrasse 11, 6330 Cham, Switzerland

Preface

Buoyancy-driven convective heat transfer from a vertical flat surface that bounds fluid-saturated porous medium is investigated with primary focus on seeking a heat transfer correlation and evidence of thermal dispersion in the vicinity of the wall. Analysis shows that dispersion caused by the tortuous pore-scale flow field can be significant in the transverse transport of energy from the wall. The present experiments suggest possible dependence of thermal dispersion on the pore-scale Peclet number in the range 1 to 10^3. However, the magnitude of transverse dispersion is smaller than that obtained for homogeneous porous media reported in the literature.

Porous media are made of water serving as the fluid phase and packed spherical beads as the solid phase. Bead materials are chosen to have either similar or dissimilar thermal conductivities compared to that of water. A medium comprising fluid and solid materials of similar thermal conductivity exhibits negligible effect of conductive heat transfer through the phases, allowing thermal dispersion to be readily observed. A medium with dissimilar thermal conductivities also allows the examination of dispersion, but the conductivity of the medium must come from published investigations. Wall temperature is employed as a non-invasive measurement technique based on the idea that the variation in time and location is an indication of flow field development adjacent to the wall. This technique with temperature sensors embedded in the wall minimizes high-frequency noise commonly observed in data taken in such a medium.

Steady-state measurements show that heat transfer can be explained by a Darcy-based model. The model assumes a velocity slip at the boundary and has no thermal dispersion. It is speculated that the reason that the model works well in the present case is that the porous medium has a lower effective Prandtl number than that of the fluid. Factors contributing to this phenomenon include the thinning of the velocity boundary layer due to flow which is restricted by shear stress from the stationary solid matrix and an increase in effective thermal conductivity due to usually higher thermal conductivity of solid than that of fluid.

The cases run for this study, which use spherical beads made of glass, polyethylene, and steel, do not exhibit significant deviation from Darcy-based studies in the literature. The present results therefore suggest possible dependence of total dispersion coefficient on the pore-scale Peclet number. The dispersion coefficient increases with Peclet number for the steel and polyethylene cases although the magnitudes are generally lower than those reported in the literature for homogeneous media. The small magnitudes may be due to high local porosity near the wall causing flow paths to be less tortuous than those away from the wall. Transient temperature profiles agree with those from conjugate analyses and confirm that the thermal conductivity of a porous medium as seen by the wall is that of the fluid-phase material. This is due to the large local porosity near the wall and also the weak dependence of the stagnant thermal conductivity on the solid-to-fluid conductivity ratio.

A review of the theory of volume averaging is included with the development of the volume-averaged energy as a foundation for determining the effective thermal conductivity of the porous medium. A model for determining thermal dispersion is also presented in connection with the present experiments.

Yokohama, Japan   Hitoshi Sakamoto
Minneapolis, MN   Francis A. Kulacki

Contents

Abbreviations

Nomenclature

A	Interface Area, m^2, Eqn. (A2)
a_L	Empirical constant for longitudinal thermal dispersion, Eqn. (2.29)
a_T	Empirical constant for transverse thermal dispersion, Eqn. (2.29)
c	Specific heat, J/kg-K
c_p	Specific heat at constant pressure, J/kg-K
$\bar{c}$	Species concentration
C	Constant
C_F	Forchheimer form drag coefficient, Eqn. (2.17)
d	Particle diameter, m
$\mathcal{D}$	Diffusion coefficient, m^2/s
D	Pore-scale dimension, m
$\mathcal{D}_{xx}$	Effective diffusion coefficient in Taylor–Aris dispersion, m^2/s, Eqn. (2.25)
f	Near-wall porosity variation, Eqn. (2.33)
$\vec{g}$	Gravitational acceleration, m/s^2
h	Heat transfer coefficient, W/m^2-K
$\bar{h}$	Average heat transfer coefficient, W/m^2-K
K	Permeability, m^2
k'	Convection-enhanced conductivity, W/m-K, Eqn. (2.1)
k_e	Effective conductivity, W/m-K, Eqn. (2.1)
k_m	Stagnant conductivity of the porous medium, W/m-K, Eqn. (2.1)
L	Characteristic length, m
L	Height of porous annulus, m
m	Parameter to define far-field temperature profile
p	Pressure, Pa
q	Heat transfer, W
q''	Heat flux, W/m^2

$\vec{r}$	Position vector, m
R	Radius, m
s	Width of porous annulus, m
T	Temperature (local or volume-averaged), K
$\widehat{T}_f$	Temperature difference between intrinsic average over fluid phase and average temperature of REV, K
$\widehat{T}_s$	Temperature difference between intrinsic average over solid phase and average temperature of REV, K
$\langle T_f \rangle$	Superficial average temperature of fluid, K
$\langle T_f \rangle^f$	Intrinsic phase average fluid temperature, K
$\langle T_s \rangle$	Superficial average temperature of solid phase, K, Eqn. (3.2)
$\langle T_s \rangle^s$	Intrinsic phase average solid temperature, K
$\widetilde{T}_f$	Temperature difference between local and intrinsic average fluid temperatures, K, Eqn. (3.20)
t	Time, s
$\vec{u}$	Velocity, (u,v,w), m/s
V	Volume, m^3
u	z-velocity, m/s
X	Non-dimensional x location
x	Longitudinal position, m
Y	Non-dimensional y location
y	Transverse position, m
Z	Non-dimensional third dimension
z	Position in the third dimension or span-wise direction, m

Greek Symbols

α_f	Thermal diffusivity of fluid, m^2/s
α_m	Thermal diffusivity of porous medium based on stagnant conductivity, k_m, m^2/s, Eqn. (2.11)
α_{dis}	Dispersivity, m^2/s, Eqn. (6.1)
α_{total}	Total dispersion coefficient, m^2/s, Eqn. (6.3)
β	Volume expansion coefficient, T^{-1}
ΔT	$T_w - T_\infty$, K
κ	Conductivity ratio, k_s/k_f
λ	Exponent for power-law temperature profile, Eqn. (2.24)
μ	Dynamic viscosity, kg/m-s
μ_e	Effective viscosity in Brinkman model, kg/m-s, Eqn. (2.15)
ν	Kinematic viscosity, m^2/s
Φ	Dissipation function, Eqn. (3.15)
ϕ	Porosity
ψ_s	Arbitrary scalar variable, Eqn. (3.7)

ρ   Density, kg/m^3
σ   Ratio of specific heats, Eqn. (2.11)
τ   Non-dimensional time, Fig. (2.7)
θ   Angle of flow in microscopic flow simulation, Eqn. (2.36)

Dimensionless Groups

Da   Darcy number, Eqn. (4.6)
Gr   Grashof number, Eqn. (4.1)
Nu   Porous medium Nusselt number, Eqn. (2.1)
Nu   Experimentally measured Nusselt number, Eqn. (4.8)
Pe   Peclet number
Pr   Prandtl number
Ra   Rayleigh number

Subscripts

0   Reference
∞   Far field
d   Pore diameter or equivalent
e   Effective, non-stagnant
f   Fluid
fs   Fluid-to-solid interface
m   Porous medium, stagnant
p   Particle
REV   Representative elementary volume
s   Solid
sf   Solid-to-fluid interface
w   Wall

Vectors and Tensors

$\overline{\overline{A}}_{dis}$ Dispersion tensor, m^2/s, Eqn. (3.41)

$\overline{\overline{A}}_{eff}$ Stagnant thermal conductivity tensor m^2/s, Eqn. (3.40)

$\vec{b}_f$ Closure variable, m

$\vec{g}$ Gravitational force vector, m/s^2

$\overline{\overline{K}}$	Permeability tensor, m^2, Eqn. (2.12)
$\overline{\overline{k}}_{tor}$	Tortuosity tensor, W/m-K, Eqn. (2.35)
$\vec{n}_{fs}$	Unit normal vector pointing from fluid to solid at solid–fluid interface, Eqn. (3.18)
$\vec{n}_{sf}$	Unit normal vector pointing from solid to fluid at solid–fluid interface, Eqn. (3.7)
$\vec{q}$	Arbitrary vector variable, Eqn. (3.8)
$\vec{u}$	Velocity vector, m/s, Eqn. (3.15)
$\overset{\sim}{\vec{u}}$	Velocity difference between local and volume-averaged fluid velocity, m/s, Eqn. (3.21)

Chapter 1
Introduction

Heat transfer in porous media has a wide range of applications. They include geothermal energy production, petroleum recovery and the storage of radioactive nuclear wastes. It was perhaps Lapwood [1], who first recognized the occurrence of buoyancy-driven convection in a packed-bed, and this coupled the heat transfer problem to fluid of flow within the medium. A porous medium can be defined as "a material consisting of a solid matrix with an interconnected void" [2]. One can imagine that the fluid mechanics within the void space can be very complicated. In general, a porous medium can be made of either a fluidized solid phase or a packed bed (stationary solid phase), can comprise either particles or fibrous solid material, can be either homogeneous or inhomogeneous, and can be either isotropic or anisotropic. Figure 1.1 is a schematic of a heterogeneous anisotropic medium with a stationary particle-like solid phase. In the interstitial voids, there may be evaporation, condensation and/or freezing of the fluid phase, as well as chemical reaction and mass transfer.

From the late 1950s through the 1960s, mathematical models were developed to explain flow within a fluid-saturated porous medium while treating it as a continuum [3–5]. The key element of this development was defining the size of an infinitesimal volume element used in the formulation of the governing transport equations that is large enough compared to the pore size. This technique today is called the method of local volume averaging. As early as the paper of Wooding [6], the method of local volume averaging was applied to the energy equation to study convective heat transfer.

In the case of free convection where there is no solid matrix restricting the flow, it is generally accepted that average Nusselt numbers for various surfaces can be expressed as power-law functions of the Rayleigh number. Such relations can be written in the form $Nu = \text{Constant} \times Ra^n$, where $Ra = g\beta(T_w - T_0)L^3/\nu\alpha_f$, and L is an appropriate length scale. The constant and exponent are generally determined from measurement and regression analysis. The Nusselt number is also scaled to the same length of the system or surface of interest, $Nu = hL/k_f$.

© The Author(s), under exclusive licence to Springer International Publishing AG,
part of Springer Nature 2018
H. Sakamoto, F. A. Kulacki, *Buoyancy-Driven Flow in Fluid-Saturated Porous
Media near a Bounding Surface*, SpringerBriefs in Applied Sciences
and Technology, https://doi.org/10.1007/978-3-319-89887-2_1

Fig. 1.1 A heterogeneous
porous medium

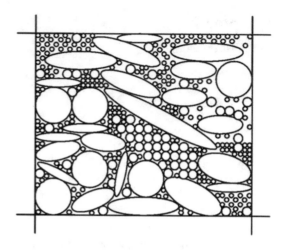

Similar correlations have been used for free convection from surfaces surrounded by fluid-saturated porous media. The literature shows that such attempts have not been very successful, especially for large Rayleigh number. When calculating the Rayleigh and Nusselt numbers, one may wonder what values are appropriate for the thermophysical properties. Certainly one would like to use the values that best represent the thermal behavior of the medium.

Figure 1.2 summarizes results of several investigations of Rayleigh-Bénard (RB) convection in a saturated porous medium [7]. The effective Nusselt numbers, based on the thermal conductivity of the stagnant porous medium, k_m, are plotted against the Rayleigh number, Ra_m, using the porous-medium thermal diffusivity, α_m, which is based on the stagnant thermal conductivity, k_m. Several investigations show that the Rayleigh number correctly predicts the onset of convection at $Ra_m \approx 40$. Measured Nusselt numbers then diverge with increasing Rayleigh number, implying that no general power-law type correlation exists between the two parameters.

Different forms of the Nusselt and Rayleigh numbers may be appropriate for free convection in porous media so as to collapse the diverging heat transfer data. A considerable amount of research has been devoted to this issue, and a great deal of disagreement exists in the literature as to how to present the results. For example, it is easier to present the Nusselt number based on the conductivity of the fluid, which shows the effect of convection relative to conduction in the fluid. What engineers would like to know however may be the effect of convection relative to conduction in the porous medium, which takes into account the properties of the solid and fluid phases. This gives rise to the notion of the stagnant thermal conductivity to be discussed later.

The Rayleigh number that may be appropriate for heat transfer in porous media can be derived from the governing differential equations by a scale analysis, i.e., $Ra_m = g\beta K L \Delta T / (\mu/\rho_0)\alpha_m$. What is specific to a porous medium is the presence of the permeability and the thermal diffusivity of the medium, α_m. Permeability, K, arises

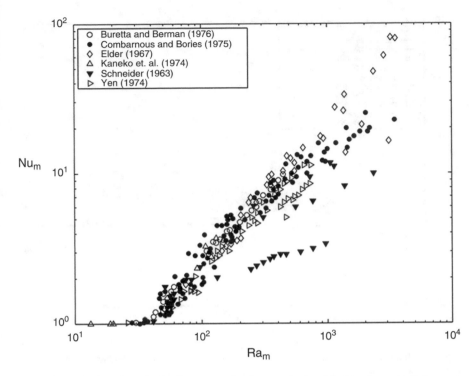

Fig. 1.2 Experimental results for free convection between horizontal plates in a saturated porous medium [7]. For these data, the Rayleigh number is defined in terms of the mean thermal diffusivity of the saturated porous medium, α_m

from Darcy's law [8], which assuming isotropy can be written $u = - (K/\mu)\partial p/\partial x$ for unidirectional flow.

The effective thermal conductivity of the porous medium is more of a concern than any of the other thermophysical properties. Experimental results (Fig. 1.2) use stagnant conductivities in the calculation of the thermal diffusivity in the Rayleigh number, although the ways in which they are calculated are different among the several investigations represented. It is speculated also that the use of non-stagnant conductivity in the Rayleigh number helps to collapse the diverging data [9]. This assumption appears reasonable because the deviations grow with increasing Rayleigh number, i.e., increasing fluid motion. However it is not clear what exactly causes the amount of deviation to be different between different experiments and how different solid-to-fluid conductivity ratios play a role. It is clear however that one needs to understand how the stagnant conductivity varies with the solid-to-fluid conductivity ratio to accurately express the thermal conductivity.

Stagnant thermal conductivities have been determined for different combinations of fluid and solid phases. Figure 1.3 shows results from several studies of heat transfer in a saturated porous medium comprising spherical beads of different sizes [7, 10–16]. It shows the stagnant conductivity of the medium normalized by the fluid

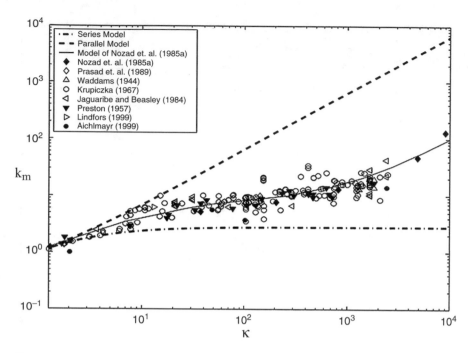

Fig. 1.3 Summary of experimental investigations for stagnant thermal conductivity of a porous medium consisting of various bead materials and fluids [7, 10–16]

conductivity as a function of the ratio of the solid-to-fluid conductivities, κ. Note that there is nearly a linear trend in the range $10 \leq \kappa \leq 1000$ on the logarithmic coordinates and relatively large scatter in the reported data over the entire range.

Our goal is to collapse the Nusselt-versus-Rayleigh data, such as shown in Fig. 1.2, and therefore develop a general correlation for natural convection from a given surface embedded in a saturated porous medium. A key to reach this goal is to successfully derive the non-stagnant conductivity as a function of parameters that explain the motion of the fluid phase. This requires the separation of motion-enhanced conductivity from the stagnant conductivity. Fluid motion and convective transport at the interface between a surface and the porous medium turn out to be important factors, and we focus our experiments on these near-wall processes. To identify what constitutes the effective thermal conductivity, volume averaging methods are extended when necessary for convective heat transfer in saturated, packed-bed porous media. As the derived governing partial differential equations are expected to be complex, model(s) may be needed for closure, and the solutions may need to be sought computationally. The geometry chosen for the present investigation is a vertical flat plate embedded in a randomly packed bed of uniform spherical particles. Only a few experimental investigations on heat transfer from such a vertical plate are available in the literature, and the research described in this monograph is intended also to contribute a fundamental technical advance.

We first review the literature on relevant sub-topics, develop the convective energy-transport equation, and report on the results of fundamental experiments. The literature review includes experimental investigations on horizontal plates that provide the data in Fig. 1.2, an investigation of stagnant conductivity that provides the best fit with experiments in Fig. 1.3, a description of fluid flow in saturated porous media in the vicinity of bounding surfaces, and recent investigations of natural convection from a vertical plate embedded in saturated porous media. As part of developing the energy transport equation, the theory of volume averaging is reviewed, which provides a firm basis for what constitutes non-stagnant thermal conductivities.

Next we describe the design and execution of measurements on heat transient heating of a vertical plate embedded in a randomly packed porous medium. These measurements and the thermal response of the plate are used to derive estimates of thermal dispersion in the near-wall region of the place surface. The overall Nusselt-versus-Rayleigh number correlation is also developed and used to validate prior analytical and numerical solutions of the governing partial differential equations. The precision and uncertainty of our measurements are included for the benefit of future research and application.

Chapter 2
Prior Research

Buoyancy-driven flow in saturated porous media has attracted many researchers from different fields because of its importance in application. The literature is broad in scope and has become massive in size, and books relevant to the present investigation have been published [2, 17, 18]. Distinct areas of interest addressed in this chapter are experimental investigations that provide Nusselt-versus-Rayleigh number correlations for heat transfer coefficients, the stagnant thermal conductivity of packed-bed porous media, the mechanics of flow through porous media in the vicinity of a bounding surface, and natural convection from a vertical plate embedded in a saturated packed-bed.

2.1 Buoyancy Driven Rayleigh-Bénard Convection

Convective heat transfer between two horizontal plates bounding a fluid is classified as the Rayleigh-Bénard (RB) problem. When the bottom plate is heated enough, for example, rotating cells develop between the plates. Similar behavior has been observed if the fluid between the plates exists within a porous medium. However complication arises even with a simple medium made of spherical beads. Experiments performed 40 years ago yet pose unresolved questions on heat transfer at large Rayleigh numbers.

Schneider [19] conducted a series of experiments using glass and steel spheres of 1.1 to 15.1 mm DIA in distilled water, turpentine oil and air in a horizontal packed bed. The stagnant conductivity, k_m, was determined by heating the top plate to create a stable temperature distribution. This made possible the calculation of the Nusselt number as a ratio of effective (non-stagnant) conductivity, k_e, to stagnant conductivity of the media, k_m,

$$\mathrm{Nu} = \frac{k_e}{k_m} = \frac{k_m + k'}{k_m} \tag{2.1}$$

© The Author(s), under exclusive licence to Springer International Publishing AG,
part of Springer Nature 2018
H. Sakamoto, F. A. Kulacki, *Buoyancy-Driven Flow in Fluid-Saturated Porous Media near a Bounding Surface*, SpringerBriefs in Applied Sciences and Technology, https://doi.org/10.1007/978-3-319-89887-2_2

where k' is the convection-enhanced component of thermal conductivity. Nusselt numbers were plotted as a function of Rayleigh number based on the stagnant conductivity, k_m. While these parameters predict the onset of convection well, the Nusselt number diverged from a single-valued solution with increasing Rayleigh number (Fig. 1.2). It was noted that smaller Nusselt numbers resulted as the ratio of the stagnant-to-fluid conductivities increased. Systems with low conductivity ratios, such as glass in water ($\kappa \approx 1.3$), followed linear trends on logarithmic coordinates implying a power-law relation between the Nusselt and Rayleigh numbers when both were based on the stagnant conductivity, k_m.

Elder [20] conducted experimental and numerical studies of convection in porous media made of spherical beads heated from below. The experiments mostly used glass spheres with 3 to 18 mm DIA. The Nusselt number as a function of Rayleigh number exhibited divergence of trend lines from a linear correlation on logarithmic coordinates. It appears that the larger the sphere diameter, the greater the deviation from the straight line. The deviation also increases with increasing Rayleigh number which is based on the stagnant conductivity. The stagnant conductivity was also used in the Nusselt number and appeared to have been obtained from the so-called parallel conduction model [21].

Kaneko et al. [22] show the effect of inclination of the heated bottom surface. The onset of convection is shifted to a lower Rayleigh number when the surface is given an angle from the horizontal direction. The Rayleigh number is based on stagnant conductivity. However, how the value is obtained is not apparent, and the value of conductivity used in the Nusselt number is not clear as well. The porous media comprised heptane, ethanol, and two sizes of silica sands. The authors noted that the shape of the sand is spherical, giving them a porosity of ~0.35 for both sizes. The results for the heptane-sand system show close agreement with those of the oil-glass system investigated by Schneider [19]. It is also noted that with relatively large solid phase conductivity, measured Nusselt numbers are lower than those for small solid phase conductivity. This result also agrees with observations made by Schneider [19].

Another experimental investigation of RB convection finds not only diverging trends in the Nusselt number with Rayleigh number but also apparent regime shifts, i.e., distinct slope changes (on logarithmic coordinates), at Rayleigh numbers between 240 and 280 [23]. At a Rayleigh number that is greater than this critical range, temperature measured at the midpoint between the top and bottom plates as a function of time becomes unstable, creating a change the observed in slope in the Nusselt-versus-Rayleigh number relation. The porous media in these experiments for Rayleigh numbers of approximately 1000 comprised a packed bed of glass and polypropylene spheres saturated with water and oil, which limits the conductivity ratio to $0.25 \leq \kappa \leq 60$ [23]. Divergent behavior of Nusselt numbers is again observed as in Fig. 1.2.

A later study hypothesizes that the divergence at high Rayleigh numbers originates from non-local thermal equilibrium within the medium and attempts to estimate the heat transfer coefficient between the phases [24]. The authors propose a mathematical model of energy transport by writing an equation for each of the two phases while including convective heat transfer between the phases,

$$\phi(\rho c)_f \frac{\partial T_f}{\partial t} + (\rho c)_f \nabla \cdot \left(\vec{u}\, T_f\right) = \nabla \cdot (k_f \nabla T_f) + h(T_s - T_f) \qquad (2.2)$$

$$(1 - \phi)(\rho c)_s \frac{\partial T_s}{\partial t} = \nabla \cdot (k_s \nabla T_s) + h(T_f - T_s) \qquad (2.3)$$

They conclude that numerical results can coincide with experimental data by adjusting the heat transfer coefficient between the fluid and solid phases [25]. As indicated by the variations in the estimated heat transfer coefficient, even for a phase conductivity ratio of unity, what governs the inter-phase heat transfer remains an issue.

Buretta and Berman [26] used glass beads of different sizes to form the solid phase in their experiments. They find a similar shift in the slope of the Nusselt number observed by Combarnous and Bia [23]. This transition was analytically determined by Gupta and Joseph [27] to be at a Rayleigh number of ~221, and Buretta and Berman [26] confirmed this prediction through their experiment (Fig. 2.1). Their experiments also show slightly diverging trends of the Nusselt number with increasing Rayleigh number. In one case, the only difference between diverging trend lines appears to be the diameter of the glass beads. As Elder [20] finds, the larger the diameter of the beads, the lower the Nusselt number at a given Rayleigh number.

Thus there appears to be some general agreement on the behavior of porous media at large Rayleigh number. The use of stagnant conductivity in the Nusselt and Rayleigh numbers is standard in the literature, although there does not appear to be agreement on how to obtain the actual value by either experimentation or modeling. These parameters prove to be a good choice in terms of predicting the onset of convection. Some studies indicate a possible regime shift in flow at a Rayleigh number of ~200 to 300; however, this transition is obscured in the Nusselt-versus-Rayleigh number data. Most importantly, lower Nusselt numbers at high Rayleigh number result when the medium consists of larger particles and/or those with large thermal conductivity [28].

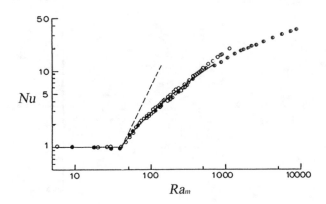

Fig. 2.1 Divergence of Nusselt numbers in layers heated from below [26]

2.2   Stagnant and Effective Thermal Conductivity

This situation leaves two issues for review at this point. The first is the stagnant thermal conductivity that may have varying effects at different Rayleigh numbers. The second is the importance of fluid flow and its effect on heat transfer. This issue is clearly suggested by the convective term in the energy equation, Eqn. (2.2). The stagnant thermal conductivity may have an important role in determining the heat transfer coefficient at high Rayleigh number. However, the stagnant thermal conductivity refers to a stationary void fluid. The use of the effective thermal conductivity is to be avoided when a medium with a stagnant fluid phase is addressed and is reserved for the more general case where the fluid phase is in motion. Many past investigations use these terms interchangeably, thus making it somewhat confusing to determine whether the fluid phase is either stationary or in motion.

Studies of stagnant conductivity seem to date back as early as Nusselt, as indicated by Jakob [29]. An extensive series of studies was conducted during the 1950s, many of which were focused on application to chemical reactors. One study shows by experiment that determination of stagnant conductivity is complicated by many factors, including the phase materials, the packing and shape of solid phase particles, and motion in the fluid phase [30]. Many models have been developed to characterize complicated heat transfer in porous media, and the simplest model may be the one-dimensional composite models, which are called the multilayer flat models [13]. These models are called the parallel and series models in more recent studies, e.g., [31], and the parallel model was used by Elder [20] as mentioned earlier.

Suppose a situation in which heat transfer is through a medium as shown in Fig. 2.2. The total heat transfer is the sum of heat transfer through each phase in parallel,

$$q = q_f + q_s \tag{2.4}$$

Fig. 2.2 The one-dimensional, parallel conduction model

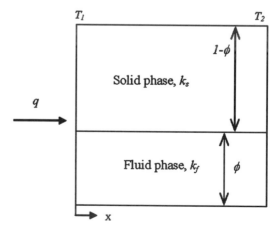

Based on Fourier's law, Eqn. (2.4), can be rewritten,

$$k_m = \phi k_f + (1 - \phi)k_s \qquad (2.5)$$

This form implies that heat is conducted through either the fluid or the solid phase with no cross transfer between them. In a similar manner, the series model can be derived,

$$\frac{1}{k_m} = \frac{1 - \phi}{k_s} + \frac{\phi}{k_f} \qquad (2.6)$$

Several experimental investigations are available that determine the stagnant thermal conductivity for various combinations of the fluid and solid phases [10–14, 32]. These experiments use spherical particles so as to simplify, i.e., homogenize, actual porous media in applications. Figure 1.3 summarizes the results of stagnant conductivities from the experiments along with the prediction by the parallel and series models [7]. The parallel model produces an upper bound estimate for relatively low ratios of the solid-to-fluid conductivity, approximately up to $\kappa = 10$. For $\kappa > 10$, the parallel model overestimates with large errors. The series model underestimates the measured thermal conductivity with increasing conductivity ratio.

Some studies make use of the parallel model. e.g., [9, 20], and one reason is that the model is implied in the two-equation model represented by Eqns. (2.2) and (2.3). These two equations can be derived from the method of volume-averaging, discussed later as it leads to a key study on stagnant conductivity. The last terms describe the effect of heat transfer between the phases, which is algebraically characterized by the temperature difference between the phases and on inter-phase heat transfer coefficient. The difficulty involved in the use of the two-equation model is to estimate the heat transfer coefficient for convection between the phases. This problem can be eliminated if the convective heat transfer can be assumed to be zero. This is to say, if the medium can be assumed to be at local thermal equilibrium, the two phases locally have the same temperature ($T_s = T_f$), and inter-phase heat transfer vanishes.

With $T_f = T_s$ in the Eqns. (2.2) and (2.3), the following one-equation model without internal energy generation is,

$$(\rho c)_m \frac{\partial T}{\partial t} + (\rho c)_f \, \vec{u} \cdot \nabla T = \nabla \cdot (k_m \nabla T) \qquad (2.7)$$

where

$$(\rho c)_m = (1 - \phi)k_s + \phi k_f \qquad (2.8)$$

$$k_m = (1 - \phi)k_s + \phi k_f \qquad (2.9)$$

The thermal conductivity of the medium used in our experiments (Chaps. 4 and 5) is the parallel conduction model (Eqn. (2.9)).

To point out one other area of disagreement in the literature, the one-equation model is further simplified, assuming constant stagnant conductivity,

$$(\rho c)_m \frac{\partial T}{\partial t} + (\rho c)_f \, \vec{u} \cdot \nabla T = k_m \nabla^2 T \tag{2.10}$$

Note that the volumetric heat capacities in the first two terms are different, i.e., one for the medium and the other for the fluid phase. It is customary in the literature on effective conductivity to define the thermal diffusivity of the medium by dividing the thermal conductivity by the volumetric heat capacity of the fluid [2, 19, 20, 22, 23, 26, 33]. Equation (2.10) then becomes,

$$\sigma \frac{\partial T}{\partial t} + \vec{u} \cdot \nabla^2 T = \alpha_m \nabla^2 T \tag{2.11}$$

where

$$\sigma = \phi + (1 - \phi) \frac{(\rho c)_s}{(\rho c)_f}$$

and $\alpha_m = k_m/(\rho c)_f$.

Nield and Bejan [2] interchangeably use this definition of the effective diffusivity, as well as a definition that divides the thermal conductivity by the specific heat of the medium, $(\rho c)_m$. When steady state heat transfer is achieved, the storage term vanishes, making it reasonable to use the specific heat of the fluid. However the thermal diffusivity of the medium appears in the Rayleigh number that is thought to be the independent variable when determining Nusselt numbers, e.g., Fig. 1.2.

Returning to the issue of stagnant conductivity, several models have been developed to characterize the spread of experimental data in Fig. 1.3, and a review article is available [31]. There are five models presented in Fig. 1.3, including the parallel and series models. For example, Batchelor and O'Brien [32] incorporate the effect of particle-to-particle contact in their model and find that the number of contact points is an important factor. This effect is increasingly important in higher conductivity ratios because much of the heat is potentially conducted through the solid phase, depending on the contact resistance. While the model predicts the stagnant conductivity quite well for $100 \leq \kappa \leq 10^3$, it does not work for low conductivity ratios.

The last two models left to be discussed in Fig. 1.3 are due to Nozad et al. [10]. The authors use the method of volume averaging to develop governing equations for the two phases and assume local thermal equilibrium to reduce to one equation. To solve the equation numerically, they assume two types of a simple two-dimensional porous medium, which are denoted the continuous and discontinuous fluid phase models. The discontinuous fluid phase model is shown in Fig. 2.3, and it assumes particle-to-particle contact on defined contact surface areas. The continuous fluid phase model lacks the arms connecting the particles. The discontinuous fluid model does a better job of predicting the data. It is reasonable to find that the discontinuous model predicts higher stagnant conductivity because of the larger solid-phase conductivity, and the

Fig. 2.3 The discontinuous fluid phase model [10]. The solid phase and contact areas are shaded. The fluid fills the open volumes

difference between experiment and theory is greater at a high conductivity ratio. The question however remains whether the models by Batchelor and O'Brien [32] and Nozad et al. [10] disagree with each other when $\kappa > 10^3$. It appears that the study by Nozad et al. is the best available to characterize stagnant conductivity of a homogeneous saturated packed bed for $\kappa \leq 10^3$. Also the method of volume averaging employed by Nozad et al. provides a rigorous theoretical approach and insight into what governs the effective thermal conductivity.

If the fluid phase is in motion, determining the effective thermal conductivity becomes more complicated, and the prediction of the Nusselt number in terms of the Rayleigh number is relatively poor (Fig. 1.2). As the Rayleigh number increases, the experimental data diverge from each other, making the Rayleigh number appear to be a poor predictor of convective heat transfer in RB convection. In an attempt to estimate effective thermal conductivity with the fluid phase in motion, we take a rigorous theoretical approach using the volume averaging method while including the convective effects in the fluid phase energy equation. Natural convection from a vertical plate will be the test case, and experiments will validate predictions.

2.3   Fluid Flow in Porous Media

Darcy [8] has been credited for his postulate of fluid motion in porous media. It is today called the Darcy's law and may be written,

$$\vec{u} = -\frac{1}{\mu}\overline{\overline{K}} \cdot \nabla p \qquad (2.12)$$

where $\overline{\overline{K}}$ is the second-order permeability tensor. For isotropic media, permeability becomes a scalar, and the above equation becomes the familiar form $\nabla p = -(\mu/K)\,\vec{u}$.

This equation is an empirical relationship that essentially defines the permeability, which is interpreted as flow conductance [18]. As is clear from Darcy's law, permeability is a function of the geometry of solid-fluid interface. Unlike the porosity, permeability explains the complex microscopic stresses exerted by the solid phase, and there is no general relationship between porosity and permeability. This has led to extensive research to determine permeability for different types of porous media. Although this topic lies beyond the scope of this monograph, the question to be investigated in our investigation has arisen in a very similar way. An important point is that Darcy's law is written today in terms of velocity and pressure gradient (intensive properties), but in the original publication of Darcy, volumetric flow rate and pressure difference as read by a manometer (extensive properties) were used. Hubbert [34] notes that the significance of changing the mathematical form lies in the fact that a manometer reading remains unchanged when the flow channel is partitioned, reducing the flow rate and area in the same proportion. This raises the question as to the velocity that one obtains by dividing flow rate by cross sectional area. The flow rate and the cross sectional area are extensive variables, whereas the ratio is an intensive or local value that varies over the cross section. Hubbert later developed the notion of the "macroscopic scale" which is better known nowadays as the representative elementary volume (REV). The REV defines the velocity as a volume-averaged quantity when Darcy's law is derived from the Navier-Stokes equation [3]. Hubbert also presents an extension of Darcy's law to include gravitational effects,

$$\frac{\mu}{K}\,\vec{u} = \rho\vec{g} - \nabla p \tag{2.13}$$

In parallel with the studies of Darcy flows, in which viscous effects on the pore scale are dominant, a departure from these flows is observed at high fluid flow rates. As the pore velocity increases, form drag starts to dominate flow resistance. An empirically derived equation for such a regime of flow may be written,

$$\nabla p = -\frac{\mu}{K}\,\vec{u} - \frac{C_F\rho_f}{\sqrt{K}}\,|\vec{u}|\vec{u} \tag{2.14}$$

where C_F is the dimensionless coefficient of the form drag. This regime of flow is often called Forchheimer flow [35]. Similar results were obtained by Dupuit [36], who studied unconfined flow through a porous medium between two reservoirs. Therefore the last term is often called the Dupuit-Forchheimer term.

Brinkman [37] suggests an extended version of Darcy's law, which includes the effect of macroscopic shear field imposed by bounding surface(s),

$$\nabla p = -\frac{\mu}{K}\,\vec{u} + \mu_e\nabla^2\vec{u} \tag{2.15}$$

where μ_e is the effective viscosity in the region with significant macroscopic shear force. Brinkman proposes this extension based on the consistency between a fluid-filled channel and a channel with a "swarm" of particles. The former case is governed by the Navier-Stokes equation, and the latter by the Darcy's law. However the Laplacian term, or the Brinkman term as it is often called, becomes significant in the vicinity of a bounding surface within a distance of $(\mu_e K/\mu)^{1/2}$, which can be derived from a scale analysis of the Brinkman equation [2].

Volume-averaging theorems have been independently derived by Slattery [4] and Whitaker [5] and demonstrate that the gradient (or divergence) of an averaged quantity is not necessarily the same as the average of the gradient (or the divergence) of the quantity (see Appendix A). These theorems have allowed the development of macroscopic governing equations for a porous medium which are essential in relating experimental measurements to the macroscopic behavior of the media.

The development of a rigorous theoretical approach to fluid motion in porous media seems to fuel debate on appropriate terms to be included in the governing equation. In an attempt to derive Darcy's law, Hubbert [3] develops an extended form that includes the transient and convection terms,

$$\rho_f \left[\frac{\partial \vec{u}}{\partial t} + \vec{u} \cdot \nabla \vec{u} \right] = -\nabla p - \frac{\mu}{K} \vec{u} \qquad (2.16)$$

This form, with or without the transient term, has been frequently used in subsequent studies (e.g., [6, 9]). A form that includes the Forchheimer term is proposed, but its validity has been questioned [38]. Following a formal volume-averaging procedure, Vafai and Tien [39] propose the following form of the momentum equation,

$$\frac{\rho}{\phi} \langle \vec{u} \cdot \nabla \vec{u} \rangle = -\nabla \langle p \rangle^f + \frac{\mu_f}{\phi} \nabla^2 \langle \vec{u} \rangle - \frac{\mu_f}{K} \langle \vec{u} \rangle + \frac{C_F \rho_f}{\sqrt{K}} |\langle \vec{u} \rangle| \langle \vec{u} \rangle \qquad (2.17)$$

where $\langle p \rangle^f$ denotes the fluid phase pressure in the REV. The empirically derived Darcy and Forchheimer terms with the surface integral term as a result of application of the volume-averaging theorem to the shear stress tensor. The inertial term is shown to be the volume-average of pointwise inertia.

Despite the controversy about the inclusion of the inertia term, the Navier-Stokes equivalent momentum equations continue to appear in the literature. Kaviany [18] introduces a form of the momentum equation which heuristically includes the transient and the body force terms,

$$\frac{\rho_f}{\phi} \left[\frac{\partial \langle \vec{u} \rangle}{\partial t} + \langle \vec{u} \rangle \cdot \nabla \langle \vec{u} \rangle \right] = -\nabla \langle p \rangle^f + \rho_f \vec{g} + \frac{\mu_f}{\phi} \nabla^2 \langle \vec{u} \rangle$$

$$\qquad (2.18)$$

$$- \frac{\mu_f}{K} \langle \vec{u} \rangle + \frac{C_f \rho_f}{\sqrt{K}} |\vec{u}| \langle \vec{u} \rangle$$

Fig. 2.4 Significant terms
in the momentum equation
as a function of Darcy and
Rayleigh numbers. (**a**)
$Pr = 1.0$. (**b**) $Pr = 0.01$.
Flow regimes: Da = Darcy,
F = Forchheimer,
CI = convective inertia,
Br = Brinkman [40]

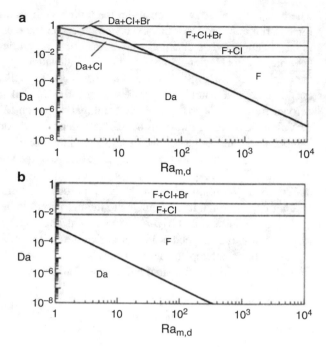

The inertia term comprises three volume-averaged velocities, which apparently contradicts the study by Vafai and Tien [39]. Lage [40] performs a scale analysis on this momentum equation and identifies significant terms as functions of Darcy and Rayleigh numbers for Prandtl numbers of 1.0 and 0.01, assuming natural convection from a vertical surface. The results are summarized in terms of Darcy number in Fig. 2.4. The Darcy number, K/H^2 is a measure of permeability of the medium with length scale H.

2.4   Natural Convection on a Vertical Plate Embedded in a Saturated Porous Medium

Only a few experimental studies are reported on natural convection on a vertical plate embedded in a saturated porous medium, which is a motivation in part for our investigation. The frequently cited first attempt at this problem is the similarity solution of Cheng and Minkowycz [41], and many analytical and numerical studies have been reported since then. At various points in all of them, effective thermal conductivities become a necessity when the fluid phase is in motion, and this problem is handled in different ways which may affect the results and prevent comparison among studies. When either an analytical or a numerical method are used, there is a missing link between a physical situation and the parameters used in them. For example, to analytically or computationally determine the dependency of the Nusselt number on the Rayleigh number, it may not be necessary to specify the values of properties that make up the Rayleigh number. Using such a study as a predictor of heat transfer coefficients will result in the difficulty when calculating the Rayleigh number, which is the key independent variable.

Cheng and Minkowycz [41] report the average Nusselt number as a function of Rayleigh number and assume that the wall temperature profile is a power-law function of distance from the leading edge. The set of governing equations used in their study resembles that used by Wooding [42]. The equations for two-dimensional flow are written with the velocities and temperature assumed to be volume-averaged quantities,

$$\frac{\partial u}{\partial x} + \frac{\partial v}{\partial y} = 0 \tag{2.19}$$

$$u = -\frac{K}{\mu_f}\left[\frac{\partial p}{\partial x} + \rho g\right] \tag{2.20}$$

$$v = -\frac{K}{\mu_f}\frac{\partial p}{\partial y} \tag{2.21}$$

$$u\frac{\partial T}{\partial x} + v\frac{\partial T}{\partial y} = \alpha_m\left(\frac{\partial^2 T}{\partial x^2} + \frac{\partial^2 T}{\partial y^2}\right) \tag{2.22}$$

The equation of sate is written in terms of the Oberbeck-Boussinesq approximation [43, 44], $\rho = \rho_0[1 - \beta(T - T_\infty)]$, which allows density variation only in the body force term so as to feed the longitudinal component of momentum and advection. Other assumptions include steady two-dimensional transport, no work, no chemical reaction or phase change, negligible viscous dissipation, negligible wall effect (no Brinkman effect), negligible inertial effect (no Forchheimer effect), and validity of Darcy's law [45]. Two additional key assumptions are the existence of local thermal equilibrium and the validity of volume-averaged quantities. The momentum equation is readily derived from the extended for of Darcy's law, and the energy equation parallels that of a pure fluid, except for the thermal diffusivity, which is that for the porous medium. The energy equation is written as a volume-averaged equation, and the variables represent average values of local quantities over of small volumetric elements containing both solid and the fluid phases. The implication therefore is that the thermal diffusivity contains all the effects of the averaging. Furthermore, the volume-averaging theorems do not directly show how the convective terms average themselves. This analysis suggests a form of the by products, which is that they are terms proportional to the second derivative of temperature. It would be convenient to contain all of these effects in the thermal diffusivity because the energy equation would be that for the fluid. On the other hand, determination of the diffusivity becomes cumbersome.

Cheng [45] suggests a form of volume-averaged energy equation in which the stagnant conductivity is separated from an added conductivity, k', due to motion of the fluid phase, which is referred to as dispersion,

$$(\rho c)_m \frac{\partial T}{\partial y} + (\rho c)_f \left[u \frac{\partial T}{\partial x} + v \frac{\partial T}{\partial y} \right] = \frac{\partial}{\partial x} \left[\left(k_m + k' \right) \frac{\partial T}{\partial x} \right] + \frac{\partial}{\partial y} \left[\left(k_m + k' \right) \frac{\partial T}{\partial y} \right]$$

(2.23)

In this way, the stagnant conductivity can be used for k_m as it is extensively researched. Details underlying dispersion are important also to the present investigation and will be discussed below.

The power-law wall temperature profile used by Cheng and Minkowycz [41] used can be written,

$$T_w = T_\infty + Cx^\lambda$$

(2.24)

where x is distance from the leading edge, measured along the wall in the direction of flow. If the exponent is zero, it becomes an isothermal profile in which the constant, C, represents the temperature difference between the wall and that of the far field. The exponent of one-third is considered the uniform heat flux case as shown via a similarity solution. A resulting Nusselt number is presented in Fig. 2.5 as a function of the exponent of the temperature profile. The Rayleigh number is based on the average temperature difference, and Nusselt number is based on the average heat

Fig. 2.5 Average Nusselt number as a function of wall temperature profile given by Eqn. (2.24). The length L is the total height of the plate [41].

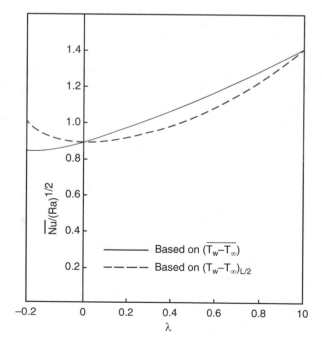

transfer coefficient. While the results are easy to use for the isothermal case ($\lambda = 0$), it is not clear how to predict the heat transfer coefficient when the power-law temperature profile is not known. Most importantly, it is assumed that the effective thermal conductivity of the medium is known a priori, which is necessary to determine the Rayleigh number.

Wooding's analysis does not contain the Brinkman term in the momentum equation, and this allows velocity slip at the vertical wall. Vafai and Tien [39] suggest the inclusion of the Brinkman and Forchheimer terms, but they note that the effects of the momentum boundary layer are important only when its thickness is larger than that of the thermal boundary layer. Nevertheless the effect of the Brinkman term is to reduce the heat transfer coefficient below that predicted by the Darcy model because of no velocity slip at the boundary. Although the Vafai-Tien numerical study was done for forced convection, the Brinkman effect is generally true in buoyancy-driven convection from a vertical wall. The question is how pronounced this effect may be in reality.

An early experimental study by Masuoka et al. [46] is reported for a vertical wall surrounding an annulus. The annulus had a height of 250 mm, and the outer wall was 328 mm in diameter and maintained at a lower temperature. The inner wall was heated and was tested with diameters of 180 mm and 280 mm to obtain different aspect ratios for the test section. Water-saturated glass beads with 2.88 mm mean diameter was the porous medium. Steady state heat transfer results were presented in terms of the Nusselt number based on the average heat transfer coefficient and stagnant thermal conductivity of the medium as functions of Rayleigh-Darcy number, RaDa, based the annular gap, s, and annulus height, H. Figure 2.6 compares

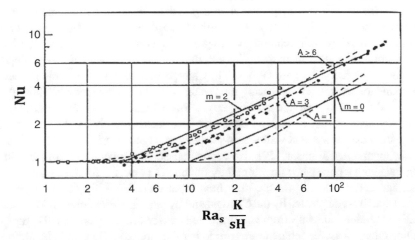

Fig. 2.6 Comparison of theory [47, 48] to measurement of heat transfer from a vertical wall of an annulus. Rayleigh and Nusselt numbers are based on the gap width of the annulus [46]. The independent variable is a product of Rayleigh number based on annular gap, s, and Darcy number based on annular gap, s, and height, H, Da = K/sH. The parameter m defines a profile of the far field temperature, $m = (T(x) - T_0)/(T_w - T_0) = m(x/H - 1/2)$, where $T_0 = T_\infty(H/2)$. A is the aspect ratio, A = H/s. Experimental data: ↑ A = 10; → A = 3.4. Numerical solution dashed line

analytical and experimental results with earlier numerical and analytical solutions [47, 48] which assumed a uniform far field temperature. The parameter, m, indicates the axial temperature gradient outside the thermal boundary layers formed on the walls, and m = 2 corresponds to the case where the free-stream temperature is the same as the wall temperature at the top of the annulus. These results clearly show the presence of the conduction regime at low Rayleigh number.

The analytical solution of Masuoka et al. [46], shown in Fig. 2.6, predicts that Nusselt numbers would be lower for a large conductivity ratio which is not the case in their experiment. They associate this effect with the porosity variation near the wall, as shown by Ofuchi and Kunii [49]. When porous media are made of spheres, local average porosity varies asymptotically from a high value at the wall to a lower value away from the wall. If the ratio of thermal conductivities is high, meaning that the solid phase has much higher conductivity than that of the fluid phase, the stagnant conductivity is lower in the vicinity of the wall. Masuoka et al. make no mention of the Brinkman effect, but both of these effects would potentially reduce heat transfer coefficients.

Imadojemu and Johnson [50] report an experimental investigation of heat transfer from a vertical plate imbedded in a porous medium comprising glass beads and water. The bead diameter of 14.6 mm gives a bulk porosity of ~0.43 and a permeability of ~4.5 × 10^{-7} m^2. The stagnant thermal conductivity is reported to be 3.9 W/mK and is used in defining the Nusselt number. The heat transfer surface is 50.8 cm (height) × 25.4 cm. Experiments were conducted with heat fluxes of 558 and 1178 W/m^2, and wall temperature was measured on the wall at 20 locations at 2.54 cm intervals. Steady heat transfer coefficients were reported in terms of local values of the Nusselt number, $Nu_x = q''x/k_m(T_w(x) - T_\infty)$ and Rayleigh-Darcy number, $Ra = g\beta x \Delta T/(\mu/\rho_0)\alpha_m$ and compared to then available analytical and numerical studies [45, 51, 52]. The authors observe nonlinear behavior of the Nusselt-versus-Rayleigh relation on logarithmic coordinates. As the Rayleigh-Darcy number, RaDa, increases, Nusselt numbers trend toward a lower slope in contrast to the predicted correlations. They suggest that this is due to the growth of the boundary layer, after comparing the results with measurements with fluid alone. At low RaDa, i.e., near the leading edge, this situation is possible because the boundary layer is thin enough so as to not be affected greatly by the solid phase, especially near the vertical wall where porosity is larger than in the mean value.

Imadojemu and Johnson [50] also report measured temperature profiles. Fitting their data with a power-law function, they obtain $\Delta T = 6.5x^{0.5}$ for the low heat input case, and $\Delta T = 10.9x^{0.5}$ for the high heat input case. Not only do the exponents differ from those suggested by the Cheng-Minkowycz similarity solution [41], actual wall temperatures deviate from the predicted power law correlations. Toward the leading edge, measured temperatures are lower than those of the Cheng-Minkowycz prediction. Toward the top of the plate, which is at the top of the annulus, measured temperatures are higher than those predicted by the power laws. This trend may have indicated the presence of stagnant warm fluid at the top of the tank, which produced a higher wall temperature. If this were the case, the exponent suggested in the predicted power law correlations would actually be lower. The exact influence of

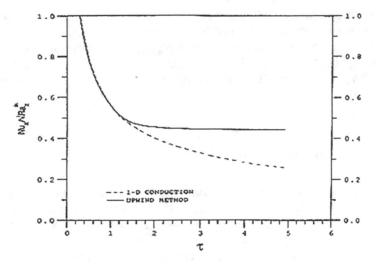

Fig. 2.7 Local heat transfer coefficient comparing conduction and upwind differencing methods of computation. Dimensionless time is $\tau = t/(K/n)$

the top region on the Nusselt number is not clear because the location of the measurements along the wall in the annulus were not reported.

Analytical and numerical studies of natural convection from a vertical wall either embedded in or bounding a porous medium are more common in the literature. The transient version of the Cheng-Minkowycz solution [41] has been studied by Ingham and Brown [53], who use a similarity solution and assume a power-law profile for the wall temperature. Haq and Mulligan [54] use an upwind numerical scheme with the added transient term and show that conduction-dominated heat transfer exists very early in the transient solution.

Figure 2.7 shows trends of the local heat transfer parameter, $Nu_x/\sqrt{Ra_x}$, as a function of the dimensionless time, $\tau = t/(K/v)$. The results of the finite-difference (upwind differencing in velocity) and one-dimensional conduction solutions are compared which indicate an early stage conduction-dominated regime. The authors emphasize that numerical diffusion is not an issue because when the steady state solution is achieved, it agrees with the Cheng-Minkowycz results. The difference in the values of the parameters by a factor of two is reasonable at large time, considering the exponent of the Rayleigh number of one half.

The majority of studies of natural convection on a vertical wall through the late 1990s involve mostly the one-equation, volume-averaged model assuming local thermal equilibrium. Since then investigations involving two-equation models appeared, e.g., the work of Rees and Pop [55]. Their model for the fluid mechanics is taken from Wooding [42]. However for heat transfer, Eqns. (2.2) and (2.3) are used. This formulation allows the phases to have different temperatures, but as noted earlier the model requires a closure condition for inter-phase heat transfer. The authors scale the inter-phase heat transfer coefficient algebraically on the

Rayleigh-Darcy number based on the thermal diffusivity of the fluid. They find larger temperature differences between the phases near the leading edge.

Once again, the major obstacle is in applying these analytical and numerical results to physical situations. It is not clear what value of the effective thermal conductivity is to be used. This situation continues to provide a challenge in even the most recent literature.

2.5 Thermal Dispersion

Dispersion in general is a term that is more commonly used today to refer to a mixing process in addition to molecular diffusion. Mixing of trace scalar quantities in the atmosphere due to turbulence has gained attention, although it is not denoted as dispersion [56]. Mixing due to turbulence has been investigated since Reynolds's pipe flow experiments in which he observed transition from laminar to turbulent flow by injecting dye [57]. These observations would later develop into extensive studies of dispersion of mass species which led to important applications in pollutant dispersion in the atmosphere and dispersion of chemical species in fixed bed reactors, to cite two areas where there is an extensive literature.

Thermal dispersion may be a more appropriate description of what causes the prediction of Nusselt numbers to diverge at high Rayleigh numbers. Using an example of tracer transport, Bear defines and explains hydrodynamic dispersion as the "...macroscopic outcome of the actual movements of the individual tracer particles through the pores and the various physical and chemical phenomena that takes place within the pores" [17, p. 580]. He states "...two basic transport phenomena involved [in hydrodynamic dispersion are] convection and molecular diffusion" [17, p. 581]. He further mentions that "...the separation between [dispersion and diffusion] is artificial [and] inseparable" [17, p. 581]. He recognizes however that molecular diffusion occurs even under no flow. A key difference between these processes may be the presence of temperature fluctuations. Many experiments have observed temperature fluctuations and deviation of the Nusselt numbers under such situations, e.g., [58, 59].

Algebraic models. Two of the early theoretical and experimental studies of mass dispersion mainly concerned longitudinal dispersion of soluble substances in a pipe flow [60, 61]. For low Reynolds numbers, axial dispersion is caused by the presence of the pipe wall. The Taylor-Aris dispersion coefficient can be defined as the effective diffusion coefficient in the diffusion equation for species c,

$$\frac{\partial \bar{c}}{\partial t} = \mathcal{D}_{xx} \frac{\partial^2 \bar{c}}{\partial x^2} \tag{2.25}$$

where $\bar{c}$ is the molar concentration. For a flow in a circular pipe at low Reynolds number,

$$\mathcal{D}_{xx} = \mathcal{D} + \frac{R^2 \bar{u}^2}{48\mathcal{D}} \tag{2.26}$$

where $\mathcal{D}$ is the molecular diffusion coefficient, R is the radius of the pipe, and $\bar{u}$ is the average longitudinal velocity. Taylor [60] experimentally confirms this analytically derived relation. Molecular diffusion can be thought of as a special case when there is no additional diffusive effect due to moving fluid. Although the effect of the additional diffusion is clearly distinguished in the Taylor-Aris model, it is difficult to experimentally separate it from molecular diffusion.

Dispersion of thermal energy in porous media has gained attention since the 1950s. The terms used initially to address the problem are *apparent* (e.g., see [12]) and *effective* thermal conductivities. Effective thermal conductivities without fluid flow in porous media have then become under extensive studies, e.g., [30]. In one of the early studies, it is noticed that, for a macroscopically isotropic porous medium, the effective thermal conductivity is different between the longitudinal and lateral directions [62]. The authors use packed spheres with airflow from the bottom at is applied from the top by an infrared lamp. The model equation is a steady state, convective diffusion equation for averaged behavior of porous medium,

$$\rho c_p \frac{\partial T}{\partial x} = k_{e,x} \frac{\partial^2 T}{\partial x^2} \tag{2.27}$$

7where $k_{e,x}$ is the effective thermal conductivity in the direction of flow. Temperature was measured at several longitudinal locations to determine the effective thermal conductivity, but it is not clear whether the measured temperatures were used directly for those in the model equation, which is the volume-averaged temperature. The experimental correlation for effective thermal conductivity in a given direction is,

$$\frac{k_{e,x}}{k_f} = \frac{k_{m,x}}{k_f} + C Pe_d \tag{2.28}$$

where $k_{m,x}$ is the stagnant conductivity of the porous medium in the x-direction, k_f is the thermal conductivity of the fluid (air in their experiment), and the Peclet number, Pe_d, is based on pore diameter, d, and the empirical constant is $0.7 < C < 0.8$. The lateral (radial) effective thermal conductivities can also be expressed in the form of Eqn. (2.28) with the empirical constant ranging from 0.1 to 0.3.

Cheng [45] suggests a separation of effective thermal conductivity into stagnant conductivity and dispersion coefficients k'_x and k'_y. A model for the dispersion coefficient is proposed [63],

$$\begin{bmatrix} \dfrac{k'_e}{k_m} \\[3mm] \dfrac{k'_y}{k_m} \end{bmatrix} = \begin{bmatrix} a_L \dfrac{C_F \sqrt{K}}{\nu} |u| \\[3mm] a_T \dfrac{C_F \sqrt{K}}{\nu} |v| \end{bmatrix} \tag{2.29}$$

where C_F is the Forchheimer coefficient, and a_L and a_T are longitudinal and transverse empirical constants. The velocities used in this model are volume-averaged quantities. Another model proposed by Plumb [64] is,

$$\begin{bmatrix} k'_x \\ k'_y \end{bmatrix} = \begin{bmatrix} 0 \\ C\rho c_p ud \end{bmatrix} \tag{2.30}$$

where C is an empirical constant, and d is the average particle diameter. The effect of dispersion in the longitudinal direction is not thought to be significant. Plumb uses this model in a system of governing equations comprising the gravity-extended Forchheimer momentum transport equation and the steady state energy equation with no longitudinal conduction. These equations are solved numerically, and results are compared with experimental data. The studies were done assuming uniform wall temperature, while experiments were conducted using uniform heat flux.

Figure 2.8 summarizes the results obtained by Plumb [64]. The Rayleigh number is based on the average pore diameter. The ordinate is normalized by a similarity solution [41] for an isothermal vertical wall. At low Rayleigh number, experimental and numerical results converge to the similarity solution. At high Rayleigh number heat transfer is enhanced by the combined effect of inertia and transverse dispersion. For the same degree of transverse dispersion, the numerical results predict a decrease in heat transfer coefficients with increasing inertia effect measured by a modified Grashof number, $Gr^* = g\beta K^{1/2} C_F(T_w - T_\infty)/v^2$.

Jiang et al. [65] use a Cheng [45, 63] type model of thermal dispersion in their numerical study which is expressed in terms of the pore diameter, d_p, and velocity components at the pore scale,

$$k' = C(\rho c)_f d_p (1 - \phi)\sqrt{u_p^2 + v_p^2} \tag{2.31}$$

where the constant $C = 1.042(\rho c)_f d_p u_p (1 - \phi)^{-0.8282}$. This model was examined in an experimentally and agrees well with results for forced convection [66].

Fig. 2.8 Effect of inertia and transverse dispersion for natural convection as a function of pore-based Rayleigh number [64]. The modified Grashof number, is $Gr^* = gbK^{1/2} C_F(T_w - T_\infty)/v^2$

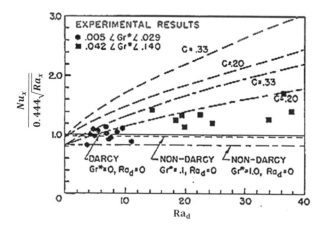

Kuo and Tien [67] propose a model for transverse thermal dispersion in forced convection where the additional component of dispersion is,

$$\frac{k'}{k_f} = CRe_p \, Pr \tag{2.32}$$

Wang and Du [68] point out that this model does not work well near a solid boundary where porosity varies. They suggest a form to include the near wall porosity variation,

$$\frac{k'}{k_f} = C(1 - \varphi)\left[fPe^{0.1} + (1 - f)Pe\right] \tag{2.33}$$

where C is an empirical constant, and f is a function of the near wall porosity variation.

After a series of radial flow experiments, Delegise et al. [69] suggest a correlation for lateral dispersion coefficient, k_{zz}, in forced convection,

$$\frac{k_{zz}}{k_e} = 0.85 + 0.0241Pe \tag{2.34}$$

Macroscopic flow analysis. Macroscopic flow models are derived from pore-scale flow analyses. The analyses are typically numerical studies of Navier-Stokes and energy equations. Kuwahara et al. [70] analytically derive an expression for thermal dispersion. They first derive a volume-averaged energy equation,

$$\left(\rho c_p\right)_f \left\langle \vec{u} \right\rangle \cdot \nabla \langle T \rangle = \nabla \cdot \left[\left(k_m \overline{\overline{I}} + \overline{\overline{k}}_{tor} + \overline{\overline{k'}}\right) \cdot \nabla \langle T \rangle \right] \tag{2.35}$$

where $\overline{\overline{K}}_{tor}$ is the tortuosity tensor and dispersion takes on tensor properties. The tortuosity and dispersion tensors are derived from the inter-phase heat transfer term that appears after the volume-averaging process. These tensors are models because the above equation assumes that the micro-scale physical processes are linearly proportional to the gradient of macroscopic temperature. A uniform temperature gradient is applied in the direction normal to the flow in an idealized porous medium (Fig. 2.9), and the normal component of the dispersion tensor is,

$$k'_{YY} = -\frac{\left(\rho c_p\right)_f / H^2}{\Delta T / H} \int_{-\frac{H}{2}}^{\frac{H}{2}} \int_{-\frac{H}{2}}^{\frac{H}{2}} (T - \langle T \rangle)\left(u - \langle u \rangle^f\right) dxdy \cdot \left(-\hat{i}\sin\theta + \hat{j}\cos\theta\right) \tag{2.36}$$

where the volume averages are oriented with the direction of flow and H is pitch distance between particles.

Simpler correlations are obtained for flow angles ranging between 15° and 45° as functions of the pore-scale Peclet number,

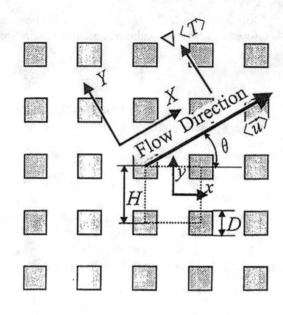

Fig. 2.9 Pore scale flow simulation in a homogeneous, anisotropic porous medium [70]. Equations (2.37) and (2.38) are valid only for the flow angles between 15° and 45°, indicating the importance of flow orientation

$$\frac{k'_{YY}}{k_f} = 0.022 \frac{Pe_d^{1.7}}{(1-\phi)^{0.25}}, \qquad Pe_d \leq 10 \tag{2.37}$$

$$\frac{k'_{YY}}{k_f} = 0.052(1-\phi)^{0.5} Pe_d, \qquad Pe_d \geq 10 \tag{2.38}$$

Kuwahara and Nakayama [71] revisit this approach and derive longitudinal dispersion coefficients whose correlation forms are,

$$\frac{k'_{YY}}{k_f} = 0.022 \frac{Pe_d^2}{(1-\phi)}, \qquad Pe_d \leq 10 \tag{2.39}$$

$$\frac{k'_{YY}}{k_f} = 2.7 \frac{Pe_d}{\phi^{0.5}}, \qquad Pe_d \geq 10 \tag{2.40}$$

It can be seen that the first correlation is equivalent to the Taylor-Aris dispersion coefficient.

2.6 Closure

All the models introduced above are based on pore-scale velocity. Local velocity must be known to estimate the effects of thermal dispersion and to implement these models for the prediction of overall heat transfer. It should also be noted that, in the case of buoyancy-driven flows from a heated non-permeable wall, transverse thermal dispersion has the potential for either increasing or decreasing overall heat transfer, depending on the magnitude of thermal dispersion [64]. The new topic presented in this monograph is based on the idea that the wall temperature variation as a function of time and location is an indication of a flow field that is developing in the vicinity of the wall.

Chapter 3
The Volume-Averaged Energy Equations

Volume averaging of the governing equations for transport in a porous medium has been a popular approach ever since Hubbert [3] conceptualized Darcy's law from the Navier-Stokes equations. The theories of area- and volume-averaging were formally developed by Slattery [4] and Whitaker [5]. Their main focus at the time was momentum transport, and energy transport was not discussed. Whitaker [72] extended his approach to develop a theory of drying in which heat, mass and momentum are transported in two-phase flow. He developed the energy transport model assuming the fluid phases, i.e., liquid and vapor, in motion; however, for closure, he assumed local thermal equilibrium. Whitaker [73] reviews a two-equation model for energy transport which does not assume local thermodynamic equilibrium, but fluid motion is assumed to be absent. In this chapter, we develop the general energy transport equations for a rigid solid phase and single-phase fluid system largely following Whitaker's approach.

3.1 The Volume-Averaged Energy Equation for the Solid Phase

The volume-averaged energy equation has been derived by Whitaker [72], and the derivation is reviewed here so that the essential features of the volume-averaging method can be pointed out. The energy equation for a solid is,

$$(\rho c)_s \frac{\partial T_s}{\partial t} = k_s \nabla^2 T_s \tag{3.1}$$

If one considers an REV greater than the typical size of the pore and less than the global domain, representative (average) behavior of variables associated with the

© The Author(s), under exclusive licence to Springer International Publishing AG, part of Springer Nature 2018
H. Sakamoto, F. A. Kulacki, *Buoyancy-Driven Flow in Fluid-Saturated Porous Media near a Bounding Surface*, SpringerBriefs in Applied Sciences and Technology, https://doi.org/10.1007/978-3-319-89887-2_3

element can be treated as a continuum. For example, the phase-average temperature can be written,

$$\langle T_s \rangle = \frac{1}{V_{REV}} \int_{V_{REV}} T_s dV \tag{3.2}$$

The phase-average temperature is often lower than the actual temperature because the solid temperature is not defined outside the solid phase. For this the intrinsic phase-average temperature is defined,

$$\langle T_s \rangle^s = \frac{1}{V_s} \int_{V_s} T_s dV \tag{3.3}$$

In a porous medium consisting of solid and fluid phases, the fraction of the volume in which the solid phase exists is $1 - \phi$, where ϕ is the local porosity. Thus the intrinsic phase-average temperature can be rewritten,

$$\langle T_s \rangle^s = \frac{1}{(1 - \phi)V_{REV}} \int_{V_{REW}} T_s dV \tag{3.4}$$

Comparing Eqns. (3.2) and (3.4),

$$\langle T_s \rangle = (1 - \phi)\langle T_s \rangle^s \tag{3.5}$$

To transform the equation for the solid phase to that for the REV, energy transfer across the phases needs to be taken into account. The volume average of Eqn. (3.1) is,

$$(\varrho c)_s \left\langle \frac{\partial T_s}{\partial t} \right\rangle = k_s \left(\left\langle \frac{\partial^2 T_s}{\partial x^2} \right\rangle + \left\langle \frac{\partial^2 T_s}{\partial y^2} \right\rangle + \left\langle \frac{\partial^2 T_s}{\partial z^2} \right\rangle \right) \tag{3.6}$$

The thermal properties have already been assumed to be constant, so they can be multiplied after the processes of averaging as shown above. At this point, the volume average of the gradient and divergence are necessary to move forward. The averaging theorem for the gradient [74] can be written,

$$\langle \nabla \psi_s \rangle = \nabla \langle \psi_s \rangle + \frac{1}{V_{REV}} \iint_{A_{sf}} \psi_s \vec{n}_{sf} dA \tag{3.7}$$

where ψ_s is the scalar to be transported, and A_{sf} is the interfacial area between the phases s and f. The normal unit vector, $\vec{n}_{sf}$ points from the s-phase to the f-phase. One of the corollaries of this theorem is the following for the divergence of a vector variable, $\vec{q}_s$,

$$\left\langle \nabla \cdot \vec{q}_s \right\rangle = \nabla \cdot \left\langle \vec{q}_s \right\rangle + \frac{1}{V_{REV}} \iint\limits_{A_{sf}} \vec{q} \cdot \vec{n}_{sf} dA \tag{3.8}$$

Finally the transient term in Eqn. (3.6) can be represented via the definition of volume-averaging, Eqn. (3.2),

$$(\rho c) \left\langle \frac{\partial T_s}{\partial t} \right\rangle = \frac{(\rho c)_s}{V_{REV}} \int\limits_{V_{REV}} \frac{\partial T_s}{\partial t} dV \tag{3.9}$$

With application of the Leibnitz formula for switching the order of the integration of a derivative, the following is the results because the integral limits are not functions of time for a rigid solid phase,

$$\frac{(\rho c)_s}{V_{REV}} \int\limits_{V_{REV}} \frac{\partial T_s}{\partial t} dV = (\rho c)_s \frac{\partial}{\partial t} \left(\frac{1}{V_{REV}} \int\limits_{V_{REV}} T_s dV \right) \tag{3.10}$$

The expression in the parentheses is the phase average temperature, so the time derivative of the temperature yields,

$$\left\langle \frac{\partial T_s}{\partial t} \right\rangle = \frac{\partial \langle T_s \rangle}{\partial t} \tag{3.11}$$

Substituting Eqns. (3.7), (3.8) and (3.11) into (3.6),

$$(\rho c)_s \frac{\partial \langle T_s \rangle}{\partial t} = k_s \left[\nabla \cdot \langle T_s \rangle + \frac{1}{V_{REV}} \iint\limits_{A_{sf}} \nabla T_s \cdot \vec{n}_{sf} dA \right]$$

$$= k_s \left\{ \nabla \cdot [\nabla \langle T_s \rangle] + \frac{1}{V_{REV}} \iint\limits_{A_{sf}} T_s \vec{n}_{sf} + \frac{1}{V_{REV}} \iint\limits_{A_{sf}} \nabla T_s \cdot \vec{n}_{sf} dA \right\} \tag{3.12}$$

Because the thermal conductivity is assumed to be constant, it can be combined with the last term to express it as heat conduction between the phases,

$$(\rho c)_s \frac{\partial \langle T_s \rangle}{\partial t} = k_s \left[\nabla^2 \langle T_s \rangle + \nabla \cdot \left(\frac{1}{V_{REV}} \iint\limits_{A_{sf}} T_s \vec{n}_{sf} dA \right) \right]$$

$$- \frac{1}{V_{REV}} \iint\limits_{A_{sf}} \vec{q}'' \cdot \vec{n}_{sf} dA \tag{3.13}$$

In Eqn. (3.13), the local conductive heat transfer is given by the Fourier's law,

$$\vec{q}'' = k_s \nabla T_s \qquad (3.14)$$

3.2   The Volume-Averaged Energy Equation for the Fluid Phase

The energy equation for an incompressible fluid phase can be written,

$$(\rho c)_f \frac{\partial T_f}{\partial t} + (\rho c)_f \nabla \cdot (\vec{u} T_f) = k_f \nabla^2 T_f + \mu_f \Phi \qquad (3.15)$$

where Φ is the dissipation function. The subscript, f, indicates liquid and distinguishes this formulation for incompressible fluid from the ideal gas formulation. The general procedure taken by Whitaker [69] can be followed. With constant properties, the volume-average of Eqn. (3.15) over the REV is,

$$(\rho c)_f \left\langle \frac{\partial T_f}{\partial t} \right\rangle + (\rho c)_f \left\langle \nabla \cdot (\vec{u}\, T_f) \right\rangle = k_f \left\langle \nabla^2 T_f \right\rangle + \mu_f \langle \Phi \rangle \qquad (3.16)$$

The Leibnitz formula can be applied to the first term,

$$\left\langle \frac{\partial T_f}{\partial t} \right\rangle = \frac{\partial \langle T_f \rangle}{\partial t} \qquad (3.17)$$

The first term on the r.h.s. of Eqn. (3.16) represents conduction. The treatment that has been made for the solid phase can also be made for the liquid phase. We next write the conduction term on the r.h.s. of Eqn. (3.13),

$$k_s \langle \nabla^2 T_f \rangle = k_f \left[\nabla^2 \langle T_f \rangle + \nabla \cdot \left(\frac{1}{V_{REV}} \iint_{A_{fs}} T_f\, \vec{n}_{fs} dA \right) \right]$$
$$- \frac{1}{V_{REV}} \iint_{A_{fs}} \vec{q}'' \cdot \vec{n}_{fs} dA \qquad (3.18)$$

where the unit normal vector, $\vec{n}_{fs}$, points from the liquid to the solid phase. Assuming no phase change, $\vec{n}_{fs} = -\vec{n}_{sf}$, and $A_{fs} = -A_{sf}$ at an arbitrary interfacial point. The last term on the r.h.s. of Eqn. (3.16) is the source term, although it is called the dissipation function. It represents the dissipation of momentum which generates heat. This term may be insignificant in low velocity flow.

The second term on the l.h.s. of Eqn. (3.16) is the convection term, and Eqn. (3.8) can be applied to rewrite the divergence,

$$\left\langle \nabla \cdot (\vec{u}\, T_f) \right\rangle = \nabla \cdot \left\langle \vec{u}\, T_f \right\rangle + \frac{1}{V_{REV}} \iint\limits_{A_{fs}} (\vec{u}\, T_f) \cdot \vec{n}_{fs} dA \qquad (3.19)$$

With no velocity slip between the phases and a rigid solid phase, the second term is zero (no convection into the solid phase). For first term on the r.h.s., decomposition [75] of transport variables is written,

$$T_f = \langle T_f \rangle^f + \tilde{T}_f \qquad (3.20)$$

$$\vec{u} = \left\langle \vec{u} \right\rangle + \tilde{\vec{u}} \qquad (3.21)$$

where the super script f applies to the fluid phase. Applying Eqns. (3.20) and (3.21) to (3.19),

$$\left\langle \nabla \cdot (\tilde{\vec{u}} T_f) \right\rangle = \nabla \cdot \left\langle \left(\left\langle \vec{u} \right\rangle + \tilde{\vec{u}} \right) \left(\langle T_f \rangle + \tilde{T}_f \right) \right\rangle$$

$$= \nabla \cdot \left\langle \left\langle \vec{u} \right\rangle \langle T_f \rangle + \nabla \cdot \left\langle \left\langle \vec{u} \right\rangle \tilde{T}_f \right\rangle + \nabla \cdot \left\langle \tilde{\vec{u}} \langle T_f \rangle \right\rangle \qquad (3.22)$$

$$+ \nabla \cdot \left\langle \tilde{\vec{u}} \tilde{T}_f \right\rangle$$

The first term on the r.h.s. is the average of the product of averaged quantities, so the outer averaging is redundant and can be eliminated. The second and the third terms are zero because the averages of fluctuating components are zero. The last term is not necessarily zero, and the averaged product of fluctuating components is called the dispersion vector [72]. Substituting Eqns. (3.17), (3.18) and (3.22) into (3.16) yields,

$$(\rho c)_f \frac{\partial \langle T_f \rangle}{\partial t} + (\rho c)_f \nabla \cdot \left(\left\langle \vec{u} \right\rangle \langle T_f \rangle \right)$$

$$= k_f \left[\nabla^2 \langle T_f \rangle + \nabla \cdot \left(\frac{1}{V_{REV}} \right) \iint\limits_{A_{fs}} T_f \vec{n}_{fs} dA \right] - (\rho c)_f \nabla \cdot \left(\tilde{\vec{u}}\, \tilde{T}_f \right) \qquad (3.23)$$

$$- \frac{1}{V_{REV}} \iint\limits_{A_{fs}} \vec{q}'' \cdot \vec{n}_{fs} dA + u_f \langle \Phi \rangle$$

3.3   Boundary Conditions

The boundary conditions at the interface between the solid and fluid phases are,

$$\vec{u}_f = 0 \tag{3.24}$$

$$\vec{q} \cdot \vec{n}_{sf} = - \vec{q} \cdot \vec{n}_{fs} \tag{3.25}$$

$$T_s = T_f \tag{3.26}$$

3.4   One-Equation Model

The solid and fluid phases can sometimes be assumed to be at local thermal equilibrium (LTE). In this case, a single temperature may be assigned to the REV. A new set of decompositions which measure deviations from LTE is,

$$\langle T_s \rangle^s = \langle T \rangle + \widehat{T}_s \tag{3.27}$$

$$\langle T_f \rangle^f = \langle T \rangle + \widehat{T}_f \tag{3.28}$$

Substituting Eqns. (3.27) and (3.28) into the volume-averaged governing equations,

$$(1 - \phi)(\rho c)_s \frac{\partial}{\partial t}\left(\langle T \rangle + \widehat{T}_s \right)$$

$$= \nabla \cdot \left\{ (1 - \phi)k_s \left[\nabla\left(\langle T \rangle + \widehat{T}_s \right) + \frac{1}{V_s} \iint_{A_{sf}} \widetilde{T}_s \vec{n}_{sf} dA \right] \right\}$$

$$- \frac{1}{V_{REV}} \iint_{A_{sf}} \vec{q}'' \, \vec{n}_{sf} dA \tag{3.29}$$

Simplifying,

$$(1-\phi)(\rho c)_s \frac{\partial \langle T \rangle}{\partial t}$$

$$= \nabla \cdot \left\{ (1-\phi)k_s \left[\nabla \langle T \rangle + \frac{1}{V_s} \iint_{A_{sf}} \widetilde{T}_s \vec{n}_{sf} dA \right] \right\} - (1-\phi)(\rho c)_s \frac{\partial \widehat{T}_s}{\partial t}$$

$$+ \nabla \cdot \left[(1-\phi)k_s \nabla \widehat{T}_s \right] - \frac{1}{V_{REV}} \iint_{A_{sf}} \vec{q}'' \cdot \vec{n}_{sf} dA$$

$$(3.30)$$

Similarly for the fluid phase,

$$\phi(\rho c)_f \frac{\partial \langle T \rangle}{\partial t} + \phi(\rho c)_f \nabla \cdot \left(\langle \vec{u} \rangle \langle T \rangle \right)$$

$$= \nabla \cdot \left\{ \phi k_f \left[\nabla \langle T \rangle + \frac{1}{V_f} \int_{A_{fs}} \widetilde{T}_f \vec{n}_{fs} dA \right] \right\}$$

$$- \phi(\rho c)_f \frac{\partial \widehat{T}_f}{\partial t} - \phi(\rho c)_f \nabla \cdot \left(\langle \vec{u} \rangle \widehat{T}_f \right) - \phi(\rho c)_f \nabla \cdot \left(\widetilde{\vec{u}} \widetilde{T}_f \right)$$

$$+ \nabla \cdot \left(\phi k_f \nabla \widehat{T}_f \right) - \frac{1}{V_{REV}} \int_{A_{fs}} \vec{q}'' \cdot \vec{n}_{fs} dA + \mu_f \langle \Phi \rangle.$$

$$(3.31)$$

Adding Eqns. (3.30) and (3.31),

$$\left[(1-\phi)(\rho c)_s + \phi(\rho c)_f \right] \frac{\partial \langle T \rangle}{\partial t} + \phi(\rho c)_f \nabla \cdot \left(\langle \vec{u} \rangle \langle T \rangle \right)$$

$$= \nabla \cdot \left\{ (1-\phi)(\rho c)_s \left[\nabla \langle T \rangle + \frac{1}{V_s} \int_{A_{sf}} \widetilde{T}_s \vec{n}_{sf} dA \right] + \phi(\rho c)_f \left[\nabla \langle T \rangle + \frac{1}{V_f} \int_{A_{fs}} \widetilde{T}_f \vec{n}_{fs} dA \right] \right\}$$

$$- (1-\phi)(\rho c)_s \frac{\partial \widehat{T}_s}{\partial t} - \phi(\rho c)_f \frac{\partial \widehat{T}_f}{\partial t} + \nabla \cdot \left[(1-\phi)k_s \nabla \widehat{T}_s \right] + \nabla \cdot \left[\phi k_f \nabla \widehat{T}_f \right]$$

$$- \phi(\rho c)_f \nabla \cdot \left(\langle \vec{u} \rangle \langle \widehat{T}_f \rangle \right) - \phi(\rho c)_f \nabla \cdot \langle \widetilde{\vec{u}} \widetilde{T}_f \rangle + \mu_f \langle \Phi \rangle.$$

$$(3.32)$$

The last three terms on the r.h.s., in addition to the macroscopic convective term, are the new terms resulting from the moving fluid phase. If fluctuating components and their derivatives are much smaller than the average values, Eqn. (3.32) further simplifies to,

$$\left[(1-\phi)(\rho c)_s + \phi(\rho c)_f \right] \frac{\partial \langle T \rangle}{\partial t} + \phi(\rho c)_f \nabla \cdot \left(\langle \vec{u} \rangle \langle T \rangle \right) =$$

$$\nabla \cdot \left\{ [(1-\phi)k_s + \phi k_f] \nabla \langle T \rangle + \frac{k_s - k_f}{V_{REV}} \iint_{A_{sf}} \widetilde{T} \vec{n}_{sf} dA \right\} - \phi(\rho c)_f \nabla \cdot \langle \widetilde{\vec{u}} \widetilde{T}_f \rangle$$

$$(3.33)$$

The transient term on the l.h.s. depends on the volumetric specific heat of the medium, which is the porosity-based average between the solid and fluid phases. The convection term depends on the property of the fluid. On the r.h.s., three distinctive terms have resulted. The first may be considered the Laplacian conduction term with the thermal conductivity is based on the parallel model. The second represents interphase conduction resulting from the difference in thermal conductivities between the solid and fluid phases. These two terms on the r.h.s. are conduction in porous medium in the absence of flow. The last term represents thermal dispersion due to fluctuating velocity and temperature.

In summary, Eqn. (3.33) has accomplished a separation of different modes of energy transport in a porous medium. It reveals the problem of stagnant thermal conductivity as the integral-differential term, which is not affected by the velocity of the fluid phase. It also reveals the problem of thermal dispersion, which creates a closure problem. This term resembles the Reynolds stress term of turbulence, and modeling will be necessary to solve the system of equations for volume-averaged temperature.

3.5   Closure

Equation (3.33) is an averaged energy equation written on the REV scale. It involves the effects of the smaller, pore scale dynamics. Specifically, the second and the third terms on the r.h.s. are desired in terms of REV-scale variables. Whitaker [73] shows a way to approach a similar closure problem for the dispersion of chemical species. A similar procedure can be followed here for energy transport. The governing equation for the fluctuating temperature component can be derived by subtracting Eqn. (3.33) from the pointwise equation for the fluid phase, Eqn. (3.15). The resulting closure problem can be stated,

$$\vec{u} \cdot \nabla \widetilde{T}_f + \widetilde{\vec{u}} \cdot \nabla \langle T_f \rangle^f = \nabla \cdot \left(\alpha_f \nabla \widetilde{T}_f \right) \tag{3.34}$$

with boundary conditions on A_{sf},

$$\vec{q} \cdot \vec{n}_{sf} = - \vec{q} \cdot \vec{n}_{fs} \tag{3.35}$$

$$\widetilde{T}_f = f(\vec{r}) \tag{3.36}$$

A scale analysis of the transient and the diffusion terms suggests that this pore-scale problem can be treated as quasi-steady. To apply this problem to an REV, Whitaker [73] suggested a periodic boundary condition while disregarding the temperature boundary condition. For the nonhomogeneous differential equation, Eqn. (3.34), a solution of the following form is suggested,

$$\widetilde{T}_f = \vec{b}_f \cdot \nabla \langle T_f \rangle^f \tag{3.37}$$

where $\vec{b}_f$ is the closure variable that maps the field of fluctuating temperature field in terms of the volume-averaged temperature. Substituting this solution into the one-equation model, Eqn. (3.33),

$$\sigma \frac{\partial \langle T \rangle}{\partial t} + \phi \nabla \cdot \left(\langle \vec{u} \rangle \langle T \rangle \right) = \nabla \cdot \left[\phi \bar{\bar{A}}_{\text{eff}} \cdot \nabla \langle T \rangle \right] + \nabla \cdot \left[\bar{\bar{A}}_{\text{dis}} \cdot \nabla \langle T \rangle \right], \tag{3.38}$$

where σ is the ratio of volumetric heat capacity,

$$\sigma = \phi + (1 - \phi) \frac{(\rho c)_s}{(\rho c)_f} \tag{3.39}$$

the stagnant thermal conductivity tensor, $\bar{\bar{A}}_{\text{eff}}$, is,

$$\bar{\bar{A}}_{\text{eff}} = \frac{(1 - \phi)k_s + \phi k_f}{(\rho c)_f} \bar{\bar{I}} + \frac{k_s - k_f}{(\rho c)_f} \frac{1}{V_{\text{REV}}} \iint_{A_{sf}} \vec{n}_{sf} \, \vec{b}_f dA \tag{3.40}$$

and the dispersion tensor is,

$$\bar{\bar{A}}_{\text{dis}} = - \left\langle \widetilde{\vec{u}} \, \vec{b}_f \right\rangle \tag{3.41}$$

The resulting energy equation appears similar to Eqn. (2.29). However, the above equation contains the components of effective conductivity and dispersion. For the case of near wall flow, the significant direction of conduction is known. Variability in the geometrical properties of porous medium is also fairly well known in the wall region. Therefore the stagnant conductivity tensor can be estimated in a straightforward manner leaving the dispersion tensor as a major challenge.

For modeling the dispersion tensor given by Eqn. (3.41), Whitaker [73] presents a governing equation for the closure variable $\vec{b}_f$. He determined that the Peclet number based on the pore dimension can be considered a measure of the significance of dispersion with respect to molecular diffusion. It is defined as the ratio of pore-scale convection (therefore dispersion) to pore-scale diffusion,

$$Pe_p = \frac{\langle u \rangle^f d_p}{\alpha_f} \frac{\phi}{1 - \phi} \tag{3.42}$$

where d_p is the particle diameter, and the second term converts the diameter to an approximate pore size [76]. Experiments conducted in homogeneous porous media show a linear increase in the ratio of thermal dispersion to molecular diffusivity for $Pe > 1$.

Chapter 4
Heat Transfer Measurements

An instrumented and heated vertical flat plate is housed in a tall cylinder filled a randomly stacked packed bed. The saturating fluid is de-aerated water. Heat transfer measurements are made when the plate is impulsively heated, and the steady-state correlation of the Nusselt-versus-Rayleigh number is determined prior to the time when the bulk temperature in the cylinder begins to rise. Thermocouples imbedded in the flat plate track the development of the thermal boundary layer. We describe the key features and thermal design elements of the apparatus as a necessary basis for interpretation and analysis of the temperatures and heat transfer measurements. Full details of the design are contained in [77].

4.1 Apparatus

The cylindrical container is 200 mm DIA × 350 mm, and a heated brass plate (230 mm × 175 mm) is inserted into it near the centerline (Figs. 4.1 and 4.2). To keep the plate design simple and compact, a symmetric assembly in which heat is allowed to leave from both sides of the plate is used. This design limits the width of the plate. The wall temperature is measured by nine embedded thermocouples at several longitudinal locations along the centerline.

The junction of each thermocouple is at the bottom of a drilled cavity approximately equal in length to the thickness of the plate so that it is as close as possible to the porous medium ($\ll$1 mm). This allows temperature measured to capture the initial conduction regime, which may be on the order of seconds depending on the heat input and thermophysical properties of the medium.

The geometry near the leading edge of the plate is similar to that used in the experiments by Imadojemu and Johnson [50]. In the present investigation, the leading edge is 50 mm above the bottom of the container. Imadojemu and Johnson

H. Sakamoto, F. A. Kulacki, *Buoyancy-Driven Flow in Fluid-Saturated Porous Media near a Bounding Surface*, SpringerBriefs in Applied Sciences and Technology, https://doi.org/10.1007/978-3-319-89887-2_4

Fig. 4.1 Cross-section of
porous medium container
[7]

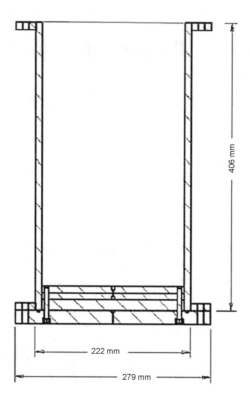

reported that the temperature toward the top of their container was always measured
to be higher than that for a power-law temperature distribution upon fitting their
experimental data. It is not clear whether such behavior was either typical of heat
transfer in porous media or due to warm stagnant fluid that rises to the top of the
container. The latter cannot be dismissed in their experiments because the gap
between the top of the container and top of the vertical plate was approximately
30 mm, and measurements are made when the system has reached steady state. In the
present study, the heated plate is designed to have approximately 76 mm between its
trailing edge and the top end of the porous medium to either eliminate or minimize
the effect of the stagnant warm fluid.

4.2 Characteristics of the Heated Plate

Various factors affect the sizing of the plate assembly. It must be tall enough to cover
a range of local Rayleigh number so that measurements of the heat transfer coefficient can be compared to the fluid-only case. A local Rayleigh number of 10^{12} is set
as the upper limit near the trailing edge (top) of the plate, which is the limit of
laminar fluid convection regime. Figure 4.3 shows estimated steady-state profiles for

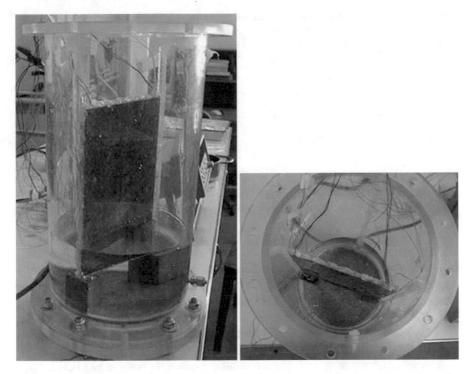

Fig. 4.2 Container with heater-plate assembly

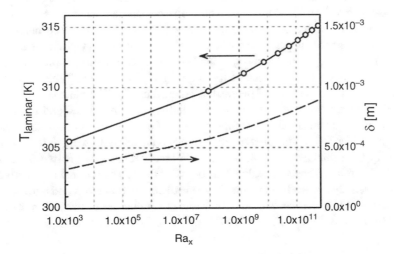

Fig. 4.3 Estimated temperature profile and the growth of the boundary layer thickness for a boundary layer flow with water and an assumed far field temperature of 300 K. The length variable in local Ra

Table 4.1 Expected Rayleigh numbers based on fluid-only convection

Fluid	Voltage (V)	Heat flux (W/m^2)	x-location (m)	Ra$_x^*$
Air	18.5	184	0.025	3.91×10^5
			0.203	1.60×10^9
Water	48.0	1290	0.025	6.61×10^6
			0.203	2.60×10^{10}
Water	98.0	5160	0.025	1.20×10^8
			0.203	4.93×10^{11}

wall temperature and boundary layer thickness for water, using the laminar flow
correlation for constant wall heat flux,

$$Nu_x = 0.60(Gr_x Pr)^{0.2} \tag{4.1}$$

where the modified local Grashof number is, $Gr_x = g\beta q_w'' x^4 / k_f \nu_f$. The correlation is
applied to the plate from the leading to the trailing edge, with the circles for the
temperatures indicating the locations of thermocouples at the local Rayleigh number,
except at the trailing edge. It should be noted that this correlation is applicable for
$Gr_x Pr > 10^5$. By this measure, all the thermocouples, except the one at the leading
edge, correspond to laminar boundary layer flow.

The upper temperature limit of the cylindrical container is set at 333 K [7], while
the plate is designed for a sufficient temperature difference between the leading and
the trailing edges to create measurable convection effects (Table 4.1). As shown in
Fig. 4.3, approximately 10 K of overall difference is expected for water. While it is
not clear a priori whether different solid phase materials cause this temperature
difference to be larger or smaller, it is expected that the difference between the
average wall temperature and the far field fluid temperature is, at least in many cases,
larger than this simple free convection analysis. The combination of plate and heater
enables experiments in the range of Rayleigh numbers shown in Table 4.1.

Figure 4.4 is a schematic of the heated plate assembly. The plate is wide enough
such that the boundary layer remains essentially two-dimensional and has a width
that is close to the inner diameter of the cylinder. Possible effects due to the presence
of the container wall on the heat transfer are minimal as well owing to the estimated
boundary layer thickness (Fig. 4.3). Alloy 260 brass is used for the two faces of the
plate with a foil-type heater sandwiched between them. The thermal conductivity of
111 W/mK at 293 K is advantageous in terms of conduction error, and a thickness of
0.0032 m (~0.125 in.) was chosen. Other relevant properties of the wall material are
(at 293 K) $\rho = 8522$ kg/m^3, c $= 385$ J/kg K, $\alpha = 3.41 \times 10^{-4}$ m^2/s.

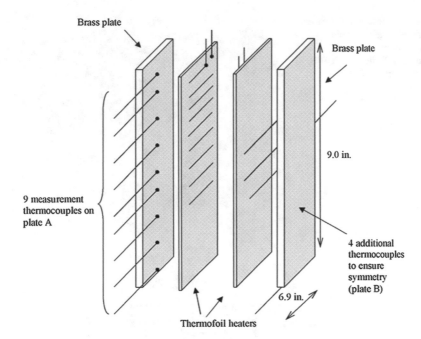

Fig. 4.4 The heater and plate assembly

4.3   Conduction Error in the Wall Temperature Measurement

The vertical wall has a finite thermal conductivity which results in an error of surface temperature measurement caused by longitudinal (x-direction) and span wise (z-direction). An additional source of error comes from the fact that the foil heating element generates heat only at the internal resistance elements that have an approximate width of 0.0032 m (~0.125 in.). The actual heat flux to the porous medium is estimated ~2.5% less than that supplied by the heaters with an uncertainty of ±2%. The analysis was done for a set of conditions that give the greatest longitudinal temperature gradient, i.e., steady state, high heat flux and water [77].

The brass wall assembly comprises a sandwiched structure of two brass plates with each having a thickness of 0.003175 m. A foil heater (Fig. 4.5) is attached to each of the plates by pre-applied adhesive from the heater manufacturer. The thickness of the heater, including the adhesive, is ~0.30 mm, a tenth of the thickness of the brass plate. Consequently, heat capacity and conduction effects of the heater are neglected. The two brass plates are glued together by a household silicone sealant, and with its relatively low thermal conductivity, heat loss is neglected. Span-wise conduction error is neglected because of nearly uniform temperature profile in this direction relative to that in the longitudinal direction. (The uniform temperature profile is part of the qualification experiment for the apparatus and is

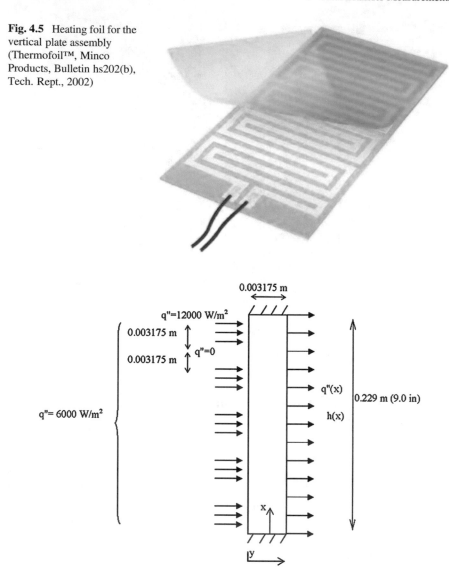

Fig. 4.5 Heating foil for the vertical plate assembly (Thermofoil™, Minco Products, Bulletin hs202(b), Tech. Rept., 2002)

Fig. 4.6 Computational domain for conduction analysis

validated in Sect. 5.1). The conduction error is expected to increase when the longitudinal temperature gradient (x-direction) becomes large, which occurs when the thermal/momentum boundary layer is fully developed in steady state and when wall heat flux is large.

As a benchmark, we assumed a wall heat flux of 6000 W/m^2 and steady-state convection in water. Figure 4.6 shows the computational domain used to estimate the heat flux. With a uniform heat flux of 6000 W/m^2 into the fluid, heat transfer coefficients are estimated with Eqn. (4.1).

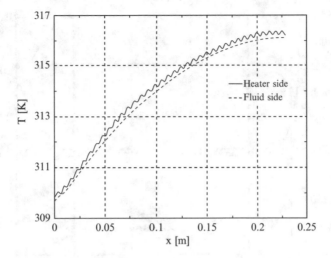

Fig. 4.7 Temperature profile of the brass plate

A 72 × 6 (x × y) grid system is generated for the numerical solution of the conduction equation for a brass plate. The 72 grid points in the longitudinal direction allow one node to have a width of the heater thereby alternating heat input to the domain. The top and the bottom of the plate are insulated boundaries, and the fluid side boundary has a prescribed distribution of heat transfer coefficient with an ambient temperature of 300 K. Figure 4.7 shows the calculated temperature profiles along the plate. The heater side of the plate has an oscillating profile because of variable heat flux supplied by the resistance elements in the heater foil. The overall temperature difference between the leading and trailing edge is a ~6 K. The wall temperature always increases from the leading edge, but the gradient becomes shallower toward the trailing edge. Temperature difference between the heater and the media sides increases with the distance from leading edge. The difference is smaller toward the leading edge because of the greater axial temperature gradient which augments axial conduction. Therefore, the error is greater toward the leading edge, as seen in Fig. 4.8 with a profile of heat flux.

Figure 4.8 shows that heat flux into the fluid is estimated to be less than 6000 W/ m^2, everywhere except the leading edge region. The error is due to the longitudinal conduction. It is always less because axial conduction is always toward the leading edge and the temperature gradient is always increasing toward it. There is an inflection point in the heat flux profile and the error profile, and this can be explained by the correspondence between decreasing heat transfer coefficient and increasing temperature difference as functions of distance from the leading edge. The maximum error occurs near the leading edge, implying that that the Nusselt correlation should not be applied in that region. The analysis must be refined to draw more accurate conclusions, or the actual heat flux into the fluid cannot be assumed to be known. In the vicinity of the leading edge, conduction dominates heat transfer from the wall to the fluid, and thus the heat flux should be less than what is estimated by the laminar

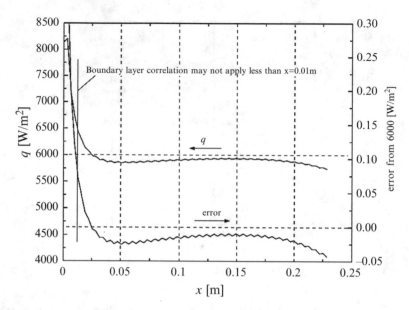

Fig. 4.8 Fluid side heat flux and error profile with alternating heat flux (brass plate with k = 111 W/mK)

boundary layer approximation. The heat flux goes to a minimum approximately at x = 0.04 m. The error is approximately 2.5%. This may be adversely affected by the leading edge region. The error remains less than 2.5% for 0.08 < x < 0.20 m. It then increases toward the trailing edge and is approximately 2% there. Based on this analysis here, we assume the corresponding uncertainty at 3%.

4.4 The Porous Medium

Spherical soda silicate glass beads with 1.5 mm and 6 mm DIA, steel beads with 6 mm and 14 mm DIA, and polyethylene balls with 25.4 mm DIA were are used for the solid phase in our experiments (Table 4.2). Air and water are readily available and easy to handle for the fluid phase, and a summary of the thermophysical properties of these materials is presented in Table 4.3. The combinations of these material provide conductivity ratios of $\kappa = 1$ (glass-water), 48 (glass-air), 100 (steel-water) and 2.4×10^3 (steel-air).

Porosity. The macroscopic volume fraction of the fluid phase (bulk porosity, ϕ) is determined by measuring the volume of water poured into the test section when the solid phase material is in place. An uncertainty in bulk porosity is expected mainly due to the effects of the surfaces in the cylindrical container. The container wall, the heated vertical wall and the supports for the vertical wall create surfaces along which the beads are aligned, creating a local porosity of 100%. This high porosity region

Table 4.2 Solid phase materials and statistical variation in diameter

Material (ϕ)	Nominal d (mm)	Average d (mm)	Std. dev. (mm)	Sample size	K (m^2) Eqn. (4.2)
Steel	6	6.00	0.00	10	4.3×10^{-8}
(0.40)	14	14.00	0.00	10	2.3×10^{-7}
Glass	1.5	1.47	0.11	14	2.1×10^{-9}
(0.38)	6	6.01	0.06	10	3.4×10^{-8}
Poly-ethylene (0.40)	25.4	25.4	0.00	10	7.7×10^{-7}

Table 4.3 Selected properties of materials for porous media

Material	Property	Source
AISI 52100 Chrome steel (1.34% Cr)	$\rho = 7865$ kg/m^3 $c = 460$ J/kg K $k = 61$ W/mK $\alpha = 1.69 \times 10^{-5}$ m^2/s	[78]
Soda silicate glass	$P = 2500$ kg/m^3 $c = 918.2$ J/kg K $k = 0.64$ W/mK $\alpha = 2.79 \times 10^{-7}$ m^2/s	[79]
High density polyethylene (HDPE)	$\rho = 958$ kg/m^3 $c = 2100$ J/kg K $k = 0.329$ W/mK $\alpha = 1.57 \times 10^{-7}$ m^2/s	[16]
Water (300 K, 1 atm)	$\rho = 996.6$ kg/m^3 $c = 4180.6$ J/kg K $k = 0.61032$ W/mK $\alpha = 1.46 \times 10^{-7}$ m^2/s	[80]
Air (293 K, 1 atm)	$P = 1.164$ kg/m^3 $c = 1012$ J/kg K $k = 0.0251$ W/mK $\alpha = 2.13 \times 10^{-5}$ mn^2/s	

near the surface extends more than a few particle diameters away from the surface. A decaying exponential function is typically used to model the variation of near-wall porosity. However, local porosity reaches a minimum of as small as 20% at a half a bead diameter away from a surface, indicating the effect of alignment of the first layer of beads [82]. The local porosity here is defined as a ratio of the area of the fluid phase to the total area at a perpendicular distance away from the impermeable wall. The local minimum porosity of 20% is smaller than a lower limit of 36% for the bulk porosity of a random packing of spherical beads measured away from a wall.

Permeability is considered flow conductance of a porous medium due to the tortuous flow paths in the Darcy flow regime. Although there is no general relationship between porosity and permeability, the literature shows that in a medium made of spherical beads, the bulk permeability can be calculated from the Kozeny formula,

$$K = \frac{d^2\phi^3}{180(1-\phi)} \tag{4.2}$$

where d is the average sphere diameter. With $d = 6$ mm and $\phi \sim 0.37$, which was consistently obtained by Kristoffersen [83], $K = 3.1 \times 10^{-8}$ m^2. Table 4.2 lists estimated permeability for the solid phases for the present measurement based on the average measured porosity for each case.

Because of the change in local porosity near an impermeable wall, it is reasonable to think that permeability changes accordingly. Rees and Pop [84] use an exponential decay function,

$$K(y) = K_\infty + (K_w - K_\infty)e^{-y/\gamma} \tag{4.3}$$

where K_∞ is far field (bulk) permeability from Eqn. (4.2), K_w is permeability at the wall, and γ is a distance parameter for the permeability variation. They find that heat transfer coefficients increase with an increase in permeability at the wall.

Forcheimer coefficient. With increasing velocity, the flow regime may depart from Darcy to Darcy-Forchheimer. The Forcheimer coefficient, C_F, represents flow resistance due to form drag caused by the presence of the solid phase. As with permeability, there is no general relation between porosity and the Forchheimer coefficient. For a porous medium made of spherical beads [85] an estimate is,

$$C_F = \frac{175(1-\phi)}{\phi^3} \tag{4.4}$$

The Forchheimer term in the momentum equation, Eqn. (2.18), is not expected to be significant, and the maximum possible flow velocity in the measurements presented in this monograph can be estimated assuming free convection in water.

Stagnant thermal conductivity. Figure 4.9 summarizes the results from experiments on stagnant conductivity as a function of the κ [7, 10–13, 15, 49, 86, 87]. For the purposes of the measurements described in this monograph, an ad hoc correlation for $10 < \kappa < 1000$ can be written,

$$\frac{k_m}{k_f} = 10^{0.4+0.3\log\kappa} \tag{4.5}$$

This range of conductivity ratios includes the glass-air ($\kappa \approx 48$) and steel-water ($\kappa \approx 100$) systems. The steel-air combination has $\kappa \approx 2430$, but it does not lie on this linear trend line.

Raleigh number and a field of independent variables. Local Rayleigh-Darcy numbers based on temperature difference and wall heat flux are,

$$(Ra_m Da)_x = \left(\frac{g\beta x^3 \Delta T}{(\mu_f/\rho_0)\alpha_m}\right)\left(\frac{K}{x^2}\right) \tag{4.6}$$

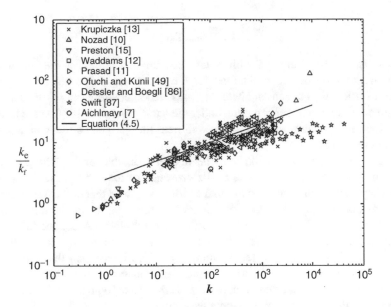

Fig. 4.9 Stagnant conductivity data and regression fit, Eqn. (4.5)

$$\left(\mathrm{Ra}_m^* \mathrm{Da}\right)_x = \left(\frac{g\beta x^4}{\nu_f \alpha_m k_m}\right)\left(\frac{K}{x^2}\right) \tag{4.7}$$

The temperature-based Rayleigh-Darcy number, Eqn. (4.6), should be as large as $\sim 10^4$ which is the approximate upper bound in the measurements of Masuoka et al. [46] (Fig. 2.6). The stagnant conductivity in the thermal diffusivity can be estimated by Eqn. (4.5). The thermal expansion coefficient, β, is taken as 2.08×10^{-4} K^{-1} for water at 293 K and 3.3×10^{-3} K^{-1} for air at 300 K [83]. With H = 0.20 m and temperature difference of 10 K, Rayleigh numbers are estimated to be 1.4×10^4, 2.9×10^3, 1.8×10^3 and 3.7×10^2 for the glass-water, glass-air, steel-air and steel-water systems, respectively. Only the glass-water system reaches the desired range of Rayleigh number.

An easy way of achieving the designed increase in Rayleigh number is to increase permeability by increasing the diameter of the spheres. If the average diameter of the spheres is 0.03 m, permeability will be K = 7.7×10^{-7} m^2, and this leads to Rayleigh numbers of 3.4×10^5, 7.2×10^4, 4.4×10^4 and 9.3×10^3 for the glass-water, glass-air, steel-air and steel-water systems, respectively.

4.5   Steady State Measurements

Measurement of wall temperatures at steady state allows the determination of local Nusselt number,

$$\text{Nu}_x = \frac{hx}{k_m} = \frac{q_w'' x}{k_m(T_x(x) - T_\infty)} \tag{4.8}$$

Results can be compared with a few key investigations in the literature. A similarity solution has been obtained for the Darcy regime in natural convection along a vertical wall at constant heat flux [41]. Another similarity solution has been obtained for the Forchheimer regime under the same conditions [88]. There is at least one experimental study with a uniform wall heat flux [50] and a few with a uniform wall temperature [46].

The wall temperature is likely to be higher than that under fluid convection because of the restricted flow due to the presence of the solid phase. However, if the solid phase has much greater thermal conductivity than that of the fluid phase, the wall temperature may not be as high. Under this scenario, the mode of heat transfer may be dominated by conduction through the solid phase. A conduction analysis using stagnant conductivity is expected to provide a limiting case for the wall temperature. The actual wall temperature should be lower than that predicted by the conduction model at all times because of enhancement of heat transfer by convection. It is also expected to increase with time so as to preserve the temperature gradient for the constant heat flux boundary condition at the wall.

4.6   Determination of the Effect of Thermal Dispersion

When the porous medium is made of glass beads and the interstitial fluid is water, the stagnant thermal conductivity of the medium is relatively easy to estimate because these two materials have similar values (Table 4.3). The parallel model for the conductivity does not cause a large error, and the effect of the non-symmetric part of the stagnant conductivity tensor in Eqn. (3.40), can be considered to be negligible.

Experimental results can be compared with analytical results from the literature. In addition to the analyses introduced in the literature review, Bejan's work [89] serves as a benchmark. A two-dimensional porous enclosure of height H and width L is heated from on the vertical boundary. The top and bottom boundaries are adiabatic, and the other vertical boundary is cooled at the same, uniform rate as the heated side. The governing equations are,

$$\frac{\partial u}{\partial y} - \frac{\partial v}{\partial x} = \frac{\rho g \beta K}{\mu} \frac{\partial T}{\partial y} \tag{4.9}$$

$$u\frac{\partial T}{\partial x} + v\frac{\partial T}{\partial x} = \alpha_m \left(\frac{\partial^2 T}{\partial x^2} + \frac{\partial^2 T}{\partial y^2} \right) \tag{4.10}$$

The first equation is a combination of the continuity and Darcy equations with a gravity extension, Eqn. (2.13). Both of these equations are assumed to be for volume averaged variables and local thermal equilibrium. The resulting Nusselt number correlation is,

$$\mathrm{Nu_x} = \frac{1}{2}\left(\frac{H}{2L}\right)^{0.2}\left(\mathrm{Ra_m^*Da}\right)_x^{0.4} \qquad (4.11)$$

The results of our experiments can be compared with this correlation, and the difference from this correlation can be assigned to the effect of thermal dispersion.

4.7   Conduction Solution

When the vertical flat plate is suddenly heated with a uniform heat flux, the adjacent porous medium initially responds by conducting the heat away. Heat transfer is approximated by one-dimensional conduction in a semi-infinite domain, and the temperature field is,

$$T(y,t) - T_i = \frac{2q_w''(\alpha_m t/\pi)^{0.5}}{k_m}\exp\left(-\frac{y^2}{4\alpha_m t}\right) - \frac{q_w''y}{k_m}\mathrm{erfc}\left(\frac{y}{2\sqrt{\alpha_m t}}\right) \qquad (4.12a)$$

where T_i is the initial temperature, α_m is based on the volumetric heat capacity of the porous medium, and the thermal conductivity and diffusivity are those of the medium in the stagnant condition. The transient wall temperature is,

$$T(0,t) - T_i = 2q_w''\left(\frac{t}{\pi(\rho c)_m k_m}\right)^{0.5} \qquad (4.12b)$$

The actual variation of the wall temperature over time is thus able to reveal the stagnant conductivity in the near-wall region.

The above conduction solution can be used to estimate the duration of experiments. It is possible that the porous medium does not reach a steady state when the mode of heat transfer is dominated by conduction. The wall temperature diverges from the solution for conduction when heat transfer reaches the acrylic container. Even if a boundary layer starts to form at this time, it is not easy to determine the far-field temperature, T_∞. Figure 4.10 shows the resulting temperature profile at $y = 5$ cm away from the vertical wall with the solid phase taken as glass. The solutions assume no convective effects, so they are the lower bound for time at which the apparatus must reach steady state. The heat fluxes of 5200 W/m^2 and 184 W/m^2 are the maximum for water and air respectively, independent of the solid phase material and particle size. Figure 4.10 shows that the steady state must be reached in ~20 min before the far field temperature experiences a significant increase.

The temperature profiles at the wall can be used to deduce the behavior of the medium. For example, Eqn. (4.12b) can be used for the initial temperature profile of the wall, assuming no temperature jump at the boundary and infinitely thin wall, or a wall with zero heat capacity (Fig. 4.11). Initial temperature gradients can be used to estimate the stagnant thermal conductivity of the medium, a valuable piece of information about the near-wall region. However, it must be noted that the porous

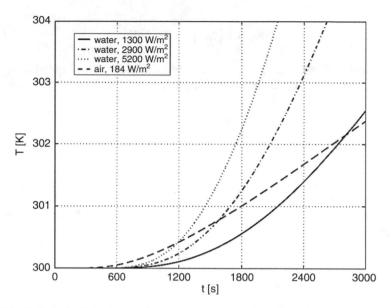

Fig. 4.10 Analytical solution for transient temperature at $y = 5$ cm, Eqn. (4.12a). The solid phase is glass

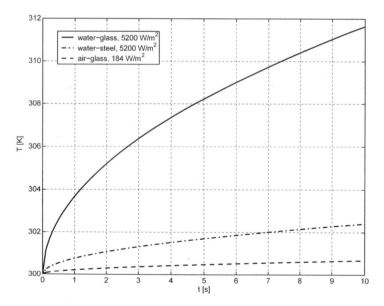

Fig. 4.11 Analytical solution for the transient wall temperature, Eqn. (4.12b). The solid phase is glass

media in our measurements are made of spherical beads. Local porosity, as well as permeability, increase toward the wall. Stagnant conductivity is also expected to change, and the value for the fluid phase material is expected to dominate. In the actual experiments, convection sets in at some time, and therefore the conduction solution is applicable only for the initial profiles. Our experiments use a wall with a finite thickness, and its properties are well known. Therefore, the temperature profile can be estimated using a conjugate heat transfer simulation.

4.8   Temperature Measurement

Owing to the sensitivity of the analysis of heat transfer measurements on wall temperatures and heat flux, considerable care is made to fully elaborate these measurements and their uncertainties. To assure precision with quantified uncertainty in temperature measurements, we calibrated all thermocouples as installed in the plate assembly. The assembly was heated in a water bath with an immersion circulator and a mercury-in-glass thermometer as the calibration standard (resolution ± 0.1 K with 0.04 K correction supplied by the manufacturer). Type E, 36 AWG Chromega-constantan thermocouples were used because of their high sensitivity (~ 80 μV/K in the range of the present study). Chromega is a chromium-nickel alloy, and the small wire diameter minimizes conduction disturbances to the plate assembly. Calibration was performed temperature range of ~ 25 to 55 °C over 3 h, and a linear regression was used to represent the temperature-emf relation. Correlation coefficients for all data fits were >0.99 with the 99% confidence interval on the order of ~ 0.17 °C. Complete calibration data are available in [77].

Chapter 5
Results

Results of a series of measurements are described and elaborated for wall temperatures and heat transfer. Owing to the sensitivity of the results to small differences in the time-temperature relations for the solid-fluid porous media, we first present details of the qualification of the heat transfer surface. Measured wall temperatures for fluid-only transient convection are then compared to measurements with the porous medium present to extract estimates of effective near wall thermal conductivity. Table 5.1 shows the experimentally determined stagnant conductivities for the solid-fluid combinations used in our measurements.

5.1 Characterization of the Heat Transfer Surface

Natural convection in water. Benchmark operation of the apparatus was based on free convection in water. This experiment was performed at three heat flux settings to cover as wide a range of Rayleigh number as possible. The resulting steady state correlation is shown in Fig. 5.1, where the Rayleigh number is based on wall heat flux $Ra_x^* = g\beta x^4 q_w'' / kv\alpha$, and an applicable correlation is Eqn. (4.1).

Transient wall conduction. The water measurements were extended to examine transient all temperatures. At the beginning of heating, the wall temperature approaches the conduction solution (Eqn. (4.12b)), and for a few seconds after the heaters are turned on, the heat capacity of the wall assembly must be taken into consideration. The result is transient, conjugate conduction that can be analyzed with a one-dimensional numerical solution technique. Figure 5.2 shows the computational domain and boundary conditions.

The water domain was assumed to be semi-infinite and the depth of the computational domain is made large enough so that its temperature remains at the initial value far from the wall throughout the simulation. The thermophysical properties

H. Sakamoto, F. A. Kulacki, *Buoyancy-Driven Flow in Fluid-Saturated Porous Media near a Bounding Surface*, SpringerBriefs in Applied Sciences and Technology, https://doi.org/10.1007/978-3-319-89887-2_5

Table 5.1 Rayleigh-Darcy numbers and experimentally estimated near wall stagnant conductivities

Fluid	Solid (d [mm])	k_m (W/m-K)	Heat flux (W/m^2)	x-location (m)	$(Ra_m^*Da)_x$
Air	Steel (14.0)	0.66	184	0.025	5.43×10^2
		0.66	184	0.203	3.48×10^4
Air	Glass (6.0)	0.17	184	0.025	2.29×10^{-1}
		0.17	184	0.203	1.47×10^1
Air	Polyethylene (25.4)	0.14	184	0.025	3.38×10^{-1}
		0.14	184	0.203	2.16×10^1
Water	Glass (1.5)	0.62	1270	0.025	7.26×10^1
		0.62	1270	0.203	4.65×10^3
		0.62	2950	0.025	1.96×10^2
		0.62	2950	0.203	1.26×10^4
		0.63	5360	0.025	4.46×10^2
		0.63	5360	0.203	2.86×10^4
Water	Glass (6.0)	0.60	77	0.025	4.14×10^1
		0.60	77	0.203	2.65×10^5
		0.60	1240	0.025	7.32×10^2
		0.60	1240	0.203	4.69×10^4
		0.61	5160	0.025	4.11×10^3
		0.61	5160	0.203	2.63×10^5
Water	Polyethylene (25.4)	0.60	1240	0.025	1.17×10^4
		0.60	1240	0.203	7.47×10^5
		0.60	2990	0.025	3.19×10^4
		0.60	2990	0.203	2.04×10^6
Water	Steel (14.0)	6.1	1260	0.025	3.00×10^1
		6.1	1260	0.203	1.92×10^3
		6.1	2820	0.025	7.71×10^1
		6.1	2820	0.203	4.93×10^3
		6.1	5190	0.025	1.53×10^2
		6.1	5190	0.203	9.79×10^3
Water	Steel (6.0)	6.1	1240	0.025	7.73×10^0
		6.1	1240	0.203	4.95×10^2
		6.1	2900	0.025	1.82×10^1
		6.1	2900	0.203	1.16×10^3
		6.1	5160	0.025	3.56×10^1
		6.1	5160	0.203	2.28×10^3

were evaluated at 310 K for water and 293 K for brass and assumed to be constant. The finite difference method employed a forward in time, space centered scheme with accuracies of first-order in time and second-order in space. The results are discussed along with the experimental results below.

Transient temperature profiles were obtained for each thermocouple individually to maximize the frequency of measurement. Measurements were made approximately every 0.036 s. Power settings of 5200 and 1250 W/m^2 were chosen to roughly correspond to the low and high settings for the experiments on the glass-water porous medium. Ten tests were conducted over 2 days for five of the nine thermocouples in the plate. The far field temperature of the water in the test apparatus

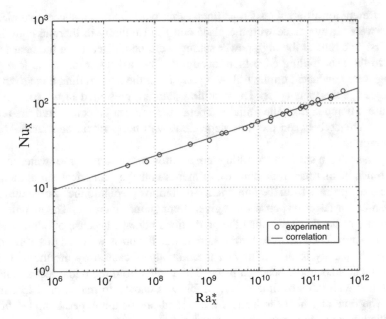

Fig. 5.1 Measured Nusselt numbers for water and the steady state heat transfer correlation, Eqn. (4.1). $Ra_x^* = g\beta x^4 q_w''/kv\alpha$

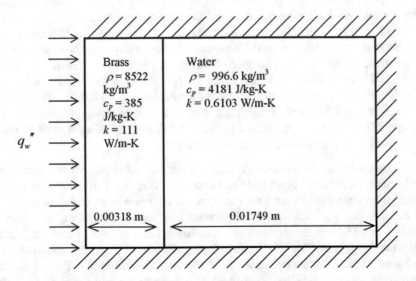

Fig. 5.2 Computational domain for one-dimensional conjugate conduction analysis (not to scale)

was different for different tests runs (increasing with time), but successive measurements were always made with the place and fluid initially in thermal equilibrium. The first five runs at the high power setting were conducted from the near leading edge to the near trailing edge thermocouples. The last five runs at the low power setting were conducted on the following day from the near trailing edge to the near leading edge thermocouples. The resulting data are presented in the form of temperature difference from the initial temperature, and can be considered the temperature difference between the time dependent wall temperature and far field water temperature.

Figures 5.3a, b show the resulting temperature trends at the brass-water interface as a function of time from both experiments and the numerical simulation. The temperature profiles from the numerical simulations predict higher temperatures at a given time, or faster response for a given temperature increase. The actual system apparently responds slower than the predictions perhaps because of additional heat capacity sandwiched between two heaters inside the brass wall and possibly smaller thermal diffusivity in the vicinity of thermocouples caused by the thermal grease fixing the thermocouple beads in place. The results from the numerical simulations are plotted with 0.5 s of time delay, which was chosen to match the experimental data. The time at which the heaters are turned on for the experimental data were identified using a sharp peak in measured voltage.

Considering these effects owing to the nature of the brass plate assembly, an apparent thermal diffusivity of the composite was calculated that best fits the temperature profiles. Figures 5.4a, b compare the experimental temperature trends with simulations done with an apparent thermal diffusivity of the brass wall of 2.7×10^{-5} m^2/s. Using this thermal diffusivity, the conjugate simulations predict the temperature profiles very well. The reason for this behavior of the plate assembly is due to water leaks in the assembly. Water has a much lower thermal diffusivity, $\sim 1.5 \times 10^{-7}$ m^2/s, than that of the brass at approximately 3.4×10^{-5} m^2/s. This hypothesis was supported by the absence of similar behavior when the measurement is made in air. Even with a small air gap in the plate assembly, the apparent thermal diffusivity was not affected because the diffusivity of air is approximately 2.1×10^{-5} m^2/s and is close to that of the brass.

Two-dimensional and transverse symmetry. Four thermocouples were embedded in the backing (secondary) brass plate to ensure two dimensionality and symmetrical behavior between the front and back surfaces of the brass plate assembly (Fig. 4.4). As an example, Fig. 5.5 shows temperature profiles at 5200 W/m^2 in water. At $x = 12.7$ and 15.2 cm, thermocouples are embedded at the center of both the primary and secondary plates. The two thermocouples at x = 14 cm, were embedded off of the center of the secondary plate to check the two-dimensionality. Two of the four thermocouples show excellent agreement with those on the primary plate. However, those at $x = 14$ and 15.2 cm show higher temperatures than the rest.

Convection in air. Measurements in air were made confirm the Nusselt number correlation, as well as to estimate an appropriate wall heat flux that results in a high enough Rayleigh number that does not cause excessively high wall temperature.

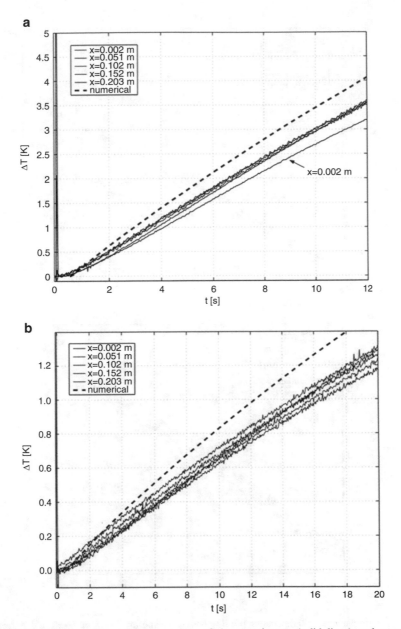

Fig. 5.3 Comparison of temperature profiles of the experiments (solid lines) and numerical simulation with uncorrected wall properties (dashed line). (**a**) High flux case, 5200 W/m². (**b**) Low flux case, 1250 W/m²

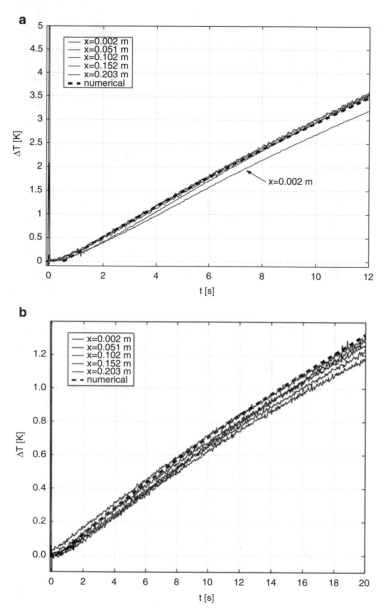

Fig. 5.4 Comparison of temperature profiles between experiments (thin lines) and numerical simulation with apparent thermal diffusivity (thick dashed line) of 2.7×10^{-5} m^2/s. (**a**) High flux, 5200 W/m^2. (**b**) Low flux, 1250 W/m^2

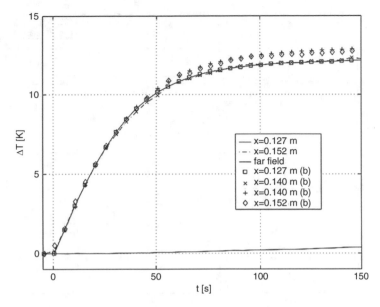

Fig. 5.5 Plate temperature trends to check two-dimensionality and transverse symmetry of the brass wall system for a heat flux of 5200 W/m^2. The x-location is in meters, and (b) indicates the secondary plate

A heat flux of 184 W/m^2 provides a Rayleigh number of 1.17 × 10^9 at the measurement point closest to the trailing edge.

Figure 5.6 presents the temperature profile as a function of x-location, along with the steady state, conjugate simulation. The far field temperature is 301 K. The conjugate simulation assumes a heat transfer coefficient profile for a constant heat flux, Eqn. (4.1), and incorporates axial conduction in the wall. The actual temperature profile results in lower temperatures than those predicted by the simulation. However, the difference is less than 1.0 K for most of the locations. The simulation reproduces the temperature gradient along the plate fairly well. The low temperature profile is suspected to be due to additional disturbance from the surrounding air and/or conduction loss from the sides of the plate assembly.

The relatively flat temperature profile suggests a large conduction error along the brass plate. Assuming air as the fluid on the convective boundary, Fig. 5.7 from the conjugate analysis shows greater axial conduction errors than the water case due to conduction in the longitudinal direction through the brass plate. The vertical wall therefore provides neither a uniform heat flux nor a uniform temperature boundary for the air, which makes porous media with air being the fluid phase difficult to analyze. It is concluded that air is not a suitable fluid at least for the apparatus of the present study.

For the initial conduction regime, wall temperatures are governed mainly by transient behavior of the brass plate and conduction into air. The problem is now spatially one-dimensional but transient. A separate conjugate analysis was

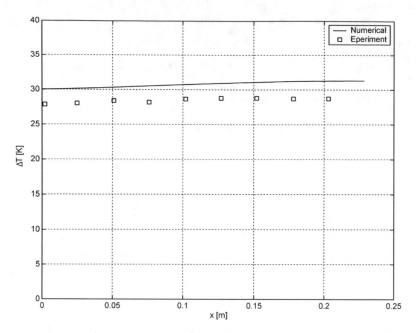

Fig. 5.6 Transient experimental data as a function of longitudinal location for air, compared with a steady state correlation. The far field temperature is 302 K, while the numerical calculation assumes 300 K

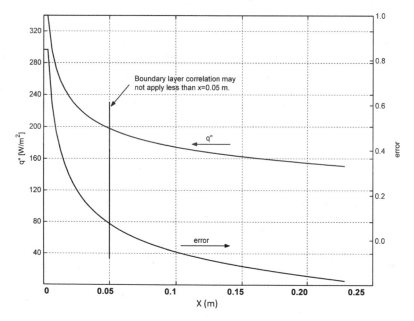

Fig. 5.7 Steady state, conjugate analysis for pure air revealing large conduction error. The error is calculated as deviation from 184 W/m²

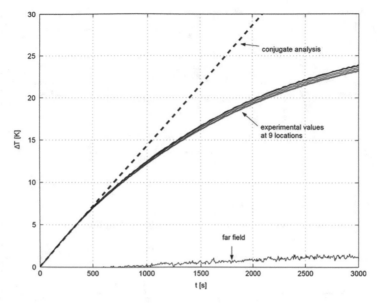

Fig. 5.8 Prediction of initial rate of temperature rise for air compared with experimental values

conducted to analytically predict the initial rate of temperature increase. The resulting initial temperature rise is approximated to be 1 K every 70 s (Fig. 5.8). The prediction appears to work well. The actual temperature profile diverges from the linear prediction because of the formation of the convective boundary layer. This is also shown by the small divergence in the actual temperature profiles among themselves.

One reason that the linear prediction appears to work well is that the heat flux into air reaches a constant value much faster than the time the system requires to depart from the conduction to the convection regime of heat transfer. Figure 5.9 shows predicted heat flux leaving the brass surface to the air from the transient conjugate simulation with a wall flux of 184 W/m^2. Heat flux is calculated by applying Fourier's law on the plate at the surface facing the convective boundary. It is zero for the initial 0.02 s, indicating the time required for heat transfer across the plate. Heat flux reaches 99% of the asymptotic maximum of 7.58 W/m^2 in ~0.23 s much sooner by at least a few hundred seconds before convective effects to appear (Fig. 5.9).

With the slowly increasing far field temperature, the system may be said to have reached a steady state. Local Nusselt numbers are plotted in Fig. 5.10 as functions of local Rayleigh number based on a wall heat flux of 184 W/m^2. The experimental correlation suggests a steeper slope than the theoretical correlation, which has an exponent of 0.20. This is perhaps due to conduction within the brass plates, making the surface more an isothermal boundary than one of constant heat flux. The average wall temperature toward the end of the run is ~302 K. The temperature difference between the leading edge and the most downstream measurement point is <1 K.

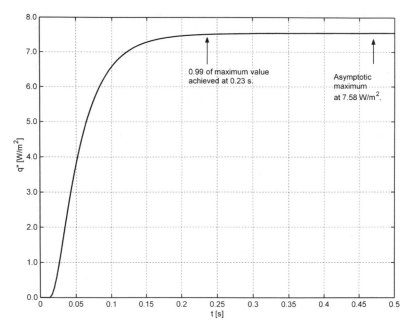

Fig. 5.9 Predicted heat flux leaving brass plate, based on temperature gradient in the brass at the surface from the conjugate transient simulation

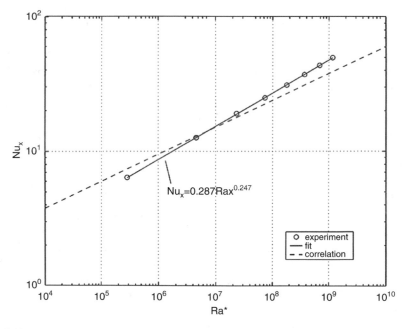

Fig. 5.10 Comparison of Nusselt number correlations between experiment and correlation for air. The change in slope is suspected to be due to axial conduction

Similar results are obtained for a lower wall heat flux of 49.2 W/m^2 with air. The temperature difference between the leading and trailing edges is limited to <0.5 K. Temperatures measured at all other locations fall between the temperatures at these two locations. The resulting Nusselt number correlation has an exponent of 0.25, while the boundary layer solution suggests 0.2 for a uniform heat flux wall in an infinite medium with a moderate Prandtl number. It is concluded at this point that air is not a suitable fluid for steady state experiments using the apparatus of the present study. For transient experiments, however, the behavior of the plate assembly is well predicted including its time response and initial rate of temperature rise.

Wall temperatures for convection in the glass-water porous medium. Glass and water have similar values for their thermal conductivities. Under thermally stable conditions, the behavior of a porous medium made of these two substances is very much like conduction through either water or glass. When instability develops, only the liquid phase begins to move. The glass beads are made of soda silicate and have a thermal conductivity of 0.64 W/mK [83]. First, transient data are examined to study the early conduction regime and near-wall behavior of the medium. Secondly, steady state cases are discussed. These cases result in heat transfer correlation at three different heat fluxes.

The porosity of the medium is estimated by measuring the volume of water contained in the tank. The estimate has yielded 0.36 for the medium used in the experiments with 6-mm glass beads. This value is consistent with other studies in the literature that use randomly packed porous media made of spherical beads and is within the theoretical range.

Transient behavior and near-wall conductivity. Figure 5.11 shows typical profiles of wall temperature as a function of time for a porous medium made of 6 mm DIA glass beads and water. The particular case presented in this figure is one for a high heat flux case of 5160 W/m^2. The data show an early conduction regime during which the increase in the wall temperature is independent of vertical position. During the experiments with a given heat flux, all thermocouple voltages are measured at a scanning period of approximately 2 s.

Data can be compared between the experiments with the glass-bead porous medium and those with pure water. Figure 5.12 compares wall temperature trends between water and the 6 mm DIA glass-water medium at $q_w'' = 5200$ W/m^2 for water and 5160 W/m^2 for the porous medium. The initial rate of temperature increase does not show significant difference between the two cases. Although the porous medium shows a higher gradient, the difference is within experimental uncertainty. Because water and glass have similar conductivity values, the stagnant conductivity in the near-wall region is not necessary to be estimated. It is expected, however, that the fluid property dominates in this region, and the conductivity of water is used as the near-wall value when calculating Rayleigh-Darcy numbers for the porous medium, for example.

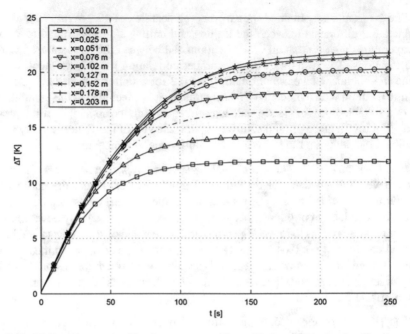

Fig. 5.11 Wall temperature profiles for a porous medium with 6 mm DIA glass beads and water at 5160 W/m^2

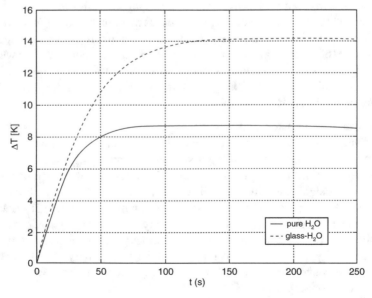

Fig. 5.12 Comparison of transient temperature profiles between pure water at 5200 W/m^2 and 6-mm glass-water medium at 5160 W/m^2, measured at $x = 0.025$ m

Fig. 5.13 Experimental
Nusselt numbers for 6-mm
glass-water medium
compared [41] and
[89]. Two lines for [89]
indicate extreme values
based on the aspect ratio of
the apparatus

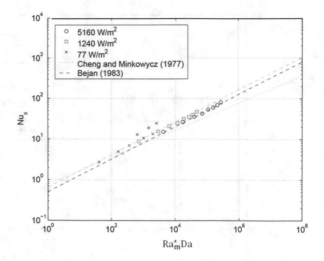

Steady-state Nusselt number correlation. Figure 5.13 shows the steady state heat
transfer data obtained with the porous medium made of 6 mm DIA glass beads and
water. These data are compared with two key analytical studies in the literature
[41, 89]. Three data points obtained at 77 W/m^2 seem to be off from a regression line
suggested by other two series of data with greater heat fluxes. These errors are due to
the very low heat flux which has resulted in maximum temperature increase of
approximately 1 K (Fig. 5.14).

The Cheng-Minkowycz [41] similarity solution can be compared with the present
results by re-calculating the temperature-based Rayleigh-Darcy number that the
authors use in their study. The correlation that they propose for a constant heat
flux boundary condition is,

$$Nu_x = 0.679(Ra_mDa)_x^{0.5} \tag{5.1}$$

where $(Ra_mDa) = \rho_\infty g\beta Kx(T_w - T_\infty)/\mu\alpha$. The flux-based Rayleigh-Darcy number
in Fig. 5.14 is $(Ra^*Da)_x = g\beta Kq_w''x^2/\alpha\nu k_m$. It follows that,

$$(RaDa)_x = \frac{(Ra^*Da)_x}{Nu_x} = (Ra^*Da)_x \frac{k(T_w - T_\infty)}{q_w''}$$

Equation (5.1) becomes,

$$Nu_x = 0.772(Ra^*Da)_x^{1/3} \tag{5.2}$$

The analytical study by Bejan [89] for a rectangular cavity uses a scale analysis
for a porous medium with a uniform heat flux on the vertical walls. Equation (4.11)
is the resulting Nusselt number correlation and is plotted in Fig. 5.13 with upper and
lower bounds based on the geometry of the present test section. A minimum aspect

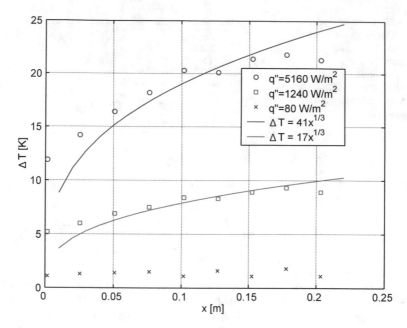

Fig. 5.14 Steady state temperature profiles as functions of longitudinal position. The power-law profiles suggested in [41] are superposed

ratio, H/L is obtained by comparing the height of the heated plate and the radius of the cylindrical container, which results in H/L = 2.1. A maximum aspect ratio of H/L = 7.6 is obtained by the ratio of the height of the container to the distance from the plate to the container wall.

It is clear that Bejan's correlation works better than that of Cheng and Minkowycz. Both studies use the same set of governing equations. The momentum equation is an extension of Darcy's law with the gravitational body force, and the energy equation uses an effective diffusivity that is assumed to be constant. The difference between the two studies is whether the medium is an enclosure [89] or semi-infinite [41]. For heat transfer in an enclosure, fluid circulates at steady state, and there is a vertical temperature gradient. Therefore the far field temperature is not constant. Bejan concludes that the wall temperature is linear, and the boundary layer thickness is constant. The present experiment does not have any cooling mechanism on the container, and the medium has a free surface exposed to the ambient temperature.

In our experiments, far field temperatures are measured approximately ~5 cm in the longitudinal and ~7 cm in the transverse positions. Unlike the water experiments, the far-field temperatures are stable at the initial values during the runs with 6 mm DIA glass beads. No significant increase is observed. This observation confirms the analysis in [89] that the medium core is motionless even in steady state.

Figure 5.14 shows steady state temperature profiles as functions of longitudinal position. Steady state temperatures are calculated as averages of the last several (typically 10 or more) measurements for each thermocouple, assuming that the system has reached steady state. The temperature differences are calculated from the average initial temperature of each thermocouple, and the plate is assumed to be initially in thermal equilibrium with the far field medium. The experimental profiles for two heat flux settings are fit with power-law profiles with an exponent of one third as suggested in [41]. The low heat flux setting of 80 W/m^2 does not develop a temperature profile that shows evidence of convection. The power-law with the one-third exponent does not appear to be a good predictor of the steady state wall temperature profiles. The profiles may be better explained by linear functions [89]. The reason that the measured profiles may look concave downward is perhaps due to conductive losses from the leading and trailing edges.

5.2   Glass-Air Porous Medium

Figure 5.15 shows temperature profiles versus time for a medium comprising 6 mm DIA glass beads and air. After more than 4 h of the test run, the system does not reach a steady state, and the far-field temperature rises at a similar rate as the wall temperature. Thus steady state data could not be expected for the glass-air medium with the present apparatus.

Conjugate heat transfer simulations are employed to compare with early temperature profiles. Figure 5.16 shows two profiles from the conjugate analysis. One corresponds to a case in which the medium is assumed to have the thermal conductivity of air, while the other corresponds to one in which the thermal conductivity is calculated based on the experimental results using spherical beads. The stagnant thermal conductivity of the glass-air system is 0.17 W/mK, and air has a conductivity of ~0.026 W/mK. The stagnant value is about 6.5 times that of air. The resulting

Fig. 5.15 Temperature profile for 6 mm DIA glass-air medium at a heat flux of 184 W/m^2

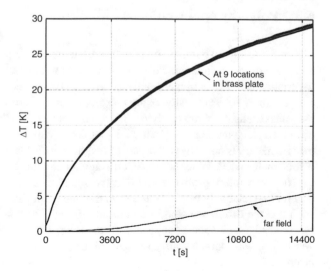

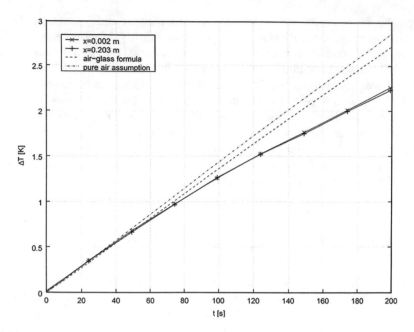

Fig. 5.16 Initial temperature profile for 6 mm DIA glass-air medium, compared with conduction solutions assuming the thermal conductivity of air in one case and that of glass-air medium. $\phi = 0.36$

temperature profiles between the two conjugate analyses yield only small differences between them before the measured temperature profiles depart from them, which indicate onset of convection.

5.3 Polyethylene-Air Porous Medium

Figure 5.17 shows a comparison of measured temperature profiles at the leading edge and the most downstream measurement location with numerically estimated profiles for conduction. The conjugate analysis presents two cases of near-wall porosity. One case assumes 100% porosity, using thermophysical properties of air. The other case assumes extension of randomly packed spherical beads to the solid surface, using an estimated stagnant thermal conductivity of 0.14 W/mK. The thermal diffusivities of the two materials differ by two orders of magnitude (Table 4.3), but this the difference is too small to determine the near-wall stagnant conductivity.

The temperature profiles show that the actual medium departs from the conduction regime relatively much sooner than the two conjugate solutions diverge enough such that the difference could be measured in an experiment. The system does not reach steady state soon enough before the outer surface of the container becomes warm. The wall remains fairly isothermal and is expected to have large conduction errors.

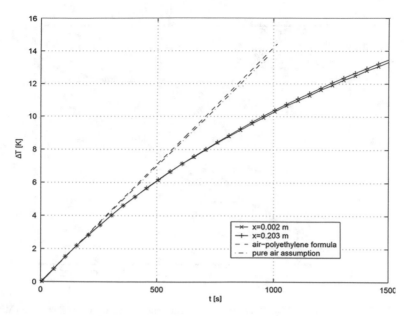

Fig. 5.17 Temperature profile for 25 mm DIA polyethylene and air at a wall heat flux of 184 W/m^2, compared with conjugate simulations for conduction

5.4   Steel-Air Porous Medium

Figure 5.18 compares experimental temperature profiles for a 14 mm DIA steel-air system with those obtained from conjugate conduction simulations. For the analysis with 36% air, a stagnant conductivity of 0.66 W/mK is used, which is ~25 times that of air. Once again, the measured temperature profiles depart from the conduction solution relatively early in the transient temperature response such that it is difficult to determine the near-wall stagnant thermal conductivity. The conjugate simulations with different thermal conductivities result in similar temperature profiles, and the near-wall conductivity cannot be estimated, i.e., the two materials have similar thermal diffusivities. Chrome steel has $\alpha = 1.69 \times 10^{-5}$ m^2/s, and the air has $\alpha = 2.13 \times 10^{-5}$ m^2/s (see Table 4.3).

5.5   Glass-Water Porous Medium

Figure 5.19 shows temperature profiles obtained from an experiment with a heat flux of 5360 W/m^2, compared with a profile estimated by a conjugate conduction simulation assuming a stagnant thermal conductivity of 0.63 W/mK. Measured temperature depart relatively early at ~20 s from the conduction solution.

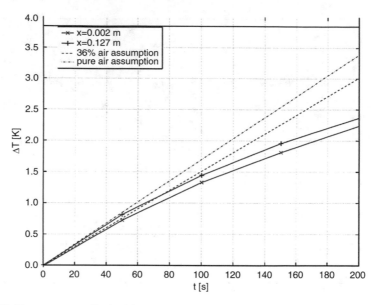

Fig. 5.18 Temperature profile for 14 mm DIA steel beads and air at a wall heat flux of 184 W/m²,
compared with conjugate simulations for the initial conduction regime

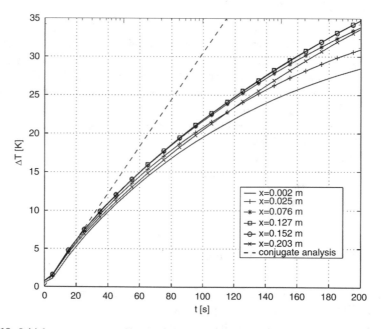

Fig. 5.19 Initial temperature profiles for 1.5-mm diameter glass-water medium at 5360 W/m²,
compared with conjugate analysis

Fig. 5.20 Experimental
Nusselt numbers for 1.5 mm
DIA glass-water medium,
compared to Cheng and
Minkowycz [41] and Bejan
[89]

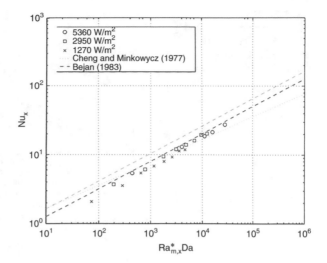

Figure 5.20 summarizes the heat transfer measurements with 1.5 mm DIA glass beads and water at three heat fluxes (5360, 2950 and 1270 W/m^2). The slope of the experimental data agrees more with Bejan's result [89] than with the Cheng-Minkowycz [41] correlation. However, there is a slight shift from the region that Bejan predicts. All experimental data are lower than the prediction, and the shift is larger for the lowest heat flux. This shift is not observed for the 6 mm DIA bead cases. For convection in water, the thickness of the boundary layer is on the order of a millimeter. Due to the Darcy effect, overall fluid motion is restricted in a porous medium. However the interaction of velocity boundary layer with the tortuous flow paths in the medium at least beyond the first layer from the vertical wall is expected more for the porous medium with 1 mm DIA beads than that with 6 mm DIA beads.

The shift in the Nusselt number may be considered due to thermal dispersion. Because of the solid matrix, there is always conductive heat transfer, largely independent of longitudinal location. Thus locations relatively far away from the wall may experience temperature increase. From a volume-averaging perspective, this is an increase in the thickness of the thermal boundary layer. As mentioned above, the velocity boundary layer may not be thicker, which may be explained by an apparent decrease in Prandtl number due to an increase in the effective thermal diffusivity.

5.6   Polyethylene-Water Porous Medium

Figure 5.21 summarizes steady state Nusselt numbers at two different heat fluxes. Again, apparent trends are shifted from the region suggested by Bejan [89]. The data closely lie along the Cheng-Minkowycz correlation; however, the slopes appear to

Fig. 5.21 Experimental
Nusselt numbers for 14-mm
DIA steel-water medium
compared to Cheng and
Minkowycz [41] and Bejan
[89]

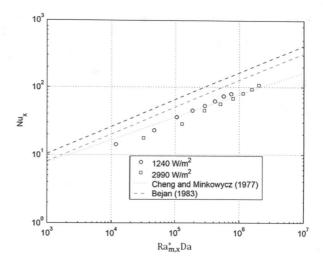

be different. The experimental data suggest a separate line for each heat flux, and the heat flux affects the lines to shift.

The shift here is not expected to be due to a decrease in apparent Prandtl number because of the low conductivity of polyethylene and the large bead diameter. The large bead diameter, on the other hand, serves to prevent them from interacting, or disturbing, the velocity boundary layer. It is possible that the velocity boundary layer here, which is ~1 mm in the absence of the beads, remains relatively two-dimensional and does not experience the Darcy effect of recirculation along the wall of the container.

5.7 Steel-Water Porous Medium

Figure 5.22 shows experimental data collected at three heat flux settings. The data are once again located below Bejan's [89] correlation. Because a similar trend is observed for the polyethylene-water system, any shift down is not solely due to thermal dispersion. Porous media with smaller diameter beads appear to have Nusselt numbers close to the Bejan correlation.

Figure 5.23 shows the results for three heat flux settings. Unlike the 14-mm DIA steel bead case, most of the data points lie inside the Bejan [89] correlation. However, the slopes are slightly different, and they are similar to that suggested by Cheng and Minkowycz [41]. Other data sets consistently resulted in an exponent of 0.4. The data for 6 mm DIA steel beads lie in the region where the Cheng-Minkowycz and Bejan correlations meet, and the small difference in the exponent makes it difficult to conclude whether the difference is significant.

Fig. 5.22 Experimental
Nusselt numbers for 6 mm
DIA steel-water medium
compared to Cheng and
Minkowycz [41] and Bejan
[89]

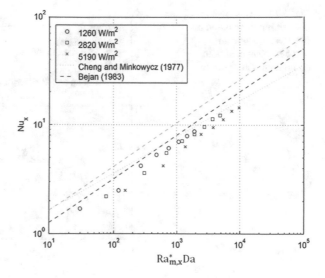

Fig. 5.23 Experimental
Nusselt numbers of all
steady state data by bead
type, compared to Cheng
and Minkowycz [41] and
Bejan [89]

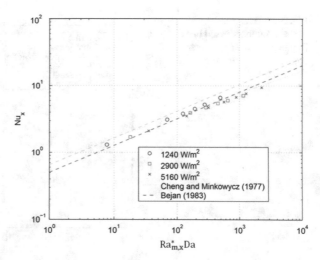

5.8   Comparison of Steady State Correlations

Figure 5.24 combines all steady state data, along with the two analytical studies.
Individual groups of data by material appear to follow Bejan's slope on logarithmic
coordinate; however, when the entire data set is examined, the data appear to follow
Cheng and Minkowycz. In terms of solid phase materials, glass cases lie most
closely to Bejan's correlation. Because the date for the 6 mm DIA steel beads lies
closely along the glass data, the shifts in 14 mm DIA steel and 25.4 mm DIA
polyethylene bead cases suggest the effects of large diameters. Figure 5.25 combines
all of the heat transfer data, where both of the parameters are based on water

Fig. 5.24 Experimental
Nusselt numbers based on
pure-water Rayleigh
number, compared with
correlation, $Nu = 0.6Ra^{0.33}$

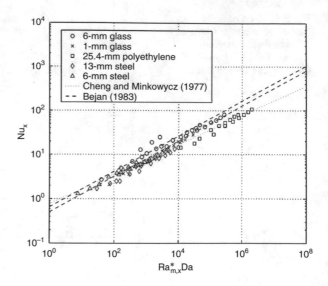

Fig. 5.25 Measured
Nusselt numbers based on
Rayleigh number for water
compared to correlation,
$Nu = 0.6Ra_x^{0.33}$

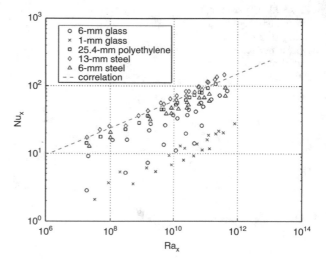

properties. The figure shows possible agreement with the water correlation,
Eqn. (4.1). The data sets that are closest to the correlation are the large diameter
sets, i.e., 14 mm DIA steel and 25.4 mm DIA polyethylene. Other sets not only shift
away from the correlation but also diverge within each set. It appears that the
diverging behavior within each set of the same material is an indication of a
significant Darcy effect. A set of data that is aligned with the correlation is for the
same heat flux setting. Sets with different fluxes are shifted in the horizontal
direction by the difference in Darcy numbers. If there is no diverging appearance,
there is no Darcy effect, and therefore heat transfer from the wall does not experience
significant effects due to the presence of the solid phase material.

Fig. 5.26 Measured
Nusselt numbers for all
steady state experiments and
theoretical results of [41, 54]

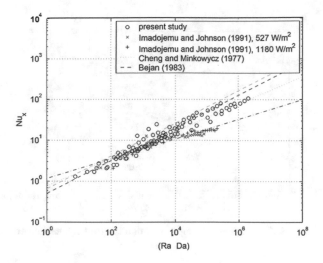

In general, the data in Fig. 5.25 show a decrease in heat transfer coefficient when compared to the benchmark case where no solid matrix is present. This is intuitive and true for most of the cases. The 14 mm DIA steel case, however, shows higher Nusselt numbers than the correlation toward the high end of Rayleigh number. This is apparently due to the high conductivity of steel. The fact that this increase only becomes significant at high Rayleigh number implies that this is augmented by convection. In other words, much heat is convected away from the vertical wall first, and there is a significant amount of heat transfer from the heated fluid to the steel beads. Furthermore, if this is the case, Nusselt number at lower Rayleigh numbers being close to the correlation indicate that the contact resistance between the beads and wall is fairly large.

Figure 5.26 compares the results of the present study with measurements of Imadojemu and Johnson [50], who conducted experiments with glass beads of a 14.6 mm DIA in water and report a best fit correlation,

$$\mathrm{Nu}_x = 1.17\left(\mathrm{Ra}_m^* \mathrm{Da}\right)_x^{0.241} \tag{5.3}$$

It appears that the only major difference is that they use beads with a larger diameter than the present investigation. The effect of bead diameter coincides with the observation that Elder [20] makes in his Rayleigh-Bénard convection experiment with glass beads.

5.9   Steady State Longitudinal Temperature Profiles

Figure 5.27 compares longitudinal temperature profiles at steady state. As briefly mentioned in Sect. 5.3, different temperature profiles are suggested in [41] and [89]. Due to experimental uncertainty, the comparison is inconclusive. However

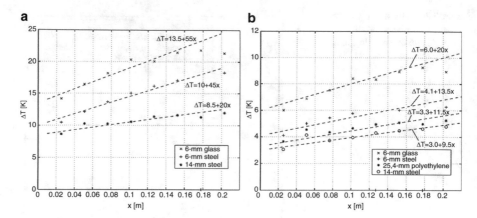

Fig. 5.27 Longitudinal temperature profiles. (**a**) High heat flux cases. (**b**) Low heat flux cases

the linear profile suggested in [89] fits the measured profiles better than power-law functions. The experimental data are fitted with linear lines, and the equations of the lines are shown. The results can be compared with Fig. 5.15, where power-law functions are used to fit heat transfer data from 6 mm DIA glass beads in water.

Chapter 6
Thermal Dispersion

6.1 Near-Wall Stagnant Thermal Conductivity

Equation (3.38) shows the effects of diffusive transport of energy as separate from those of dispersive transport, and Eqn. (3.40) defines the stagnant conductivity tensor and shows that it is a function of the geometry of the porous medium and is not affected by the presence of fluid motion. Measured stagnant conductivities have been reported extensively in the literature and are summarized in Fig. 4.9. Some difficulties are associated with an attempt to apply these results to the present investigation. One is that the experimental results do not appear to be explained by a simple equation. Even if an equation is fit by regression, the data points have relatively large deviations from each other, possibly caused by the relatively large experimental uncertainty due to the spread of the experimental data points. Secondly and more importantly, these experiments have used measurement devices for which the characteristic dimension is much larger than particle diameter so that the local porosity variation near the solid wall can be neglected. These experiments report a bulk porosity of 36–37%, indicating randomly packed spherical beads. When the beads are packed, they self-assemble along an impermeable solid wall creating high porosity at the wall. The local porosity as a function of distance away from the wall quickly decreases to a minimum ~20%, much lower than the average (bulk) value, and the average porosity reaches the bulk value several bead diameters away from the wall. In the present study, the distance from the wall that is significant for thermal diffusion is expected to be on the order of, or less than, the distance of the porosity variation. Within one particle radius from the wall, for example, the stagnant conductivity is dominated by the value for the fluid phase.

Near-wall stagnant conductivities are estimated using initial transient temperature profiles obtained in our experiments. Figure 6.1 shows early time temperature profiles with 6 mm DIA steel beads and water, along with three estimates using different near-wall conductivity values. A conductivity of 6.1 W/mK is based on Eqn. (4.5),

© The Author(s), under exclusive licence to Springer International Publishing AG, part of Springer Nature 2018
H. Sakamoto, F. A. Kulacki, *Buoyancy-Driven Flow in Fluid-Saturated Porous Media near a Bounding Surface*, SpringerBriefs in Applied Sciences and Technology, https://doi.org/10.1007/978-3-319-89887-2_6

Fig. 6.1 Early time
temperature profiles for
6-mm DIA steel beads in
water. Predicted profiles are
from conjugate conduction
simulation with three values
of estimated conductivity
values. The conjugate
solutions are adjusted by 2 s
to incorporate initial warm-
up of the wall

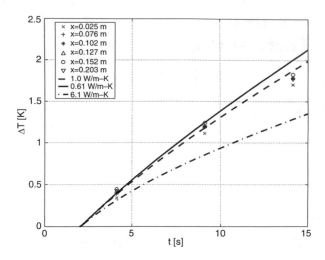

Table 6.1 Near-wall thermal
conductivities, $k_{m,w}$

Porous medium	$k_{m,w}/k_f$
6 mm DIA glass-water	1
1.5 mm DIA glass-water	1
25.4 mm DIA polyethylene-water	1
14 mm DIA steel-water	1.6
6 mm DIA steel-water	1.6

assuming random packing with $\phi = 0.36$. This value does not apparently apply in
the vicinity of the wall. An extreme case is to assume $\phi = 1$, in which case the
conductivity value of the fluid (water) applies. It captures the slope of the experimental data much better than the random packing assumption. The best estimate for
this case of the steel-water medium is 1.0 W/mK. The slight shift between the
experimental data and the prediction is due to the difficulty associated with locating
the true time zero.

Table 6.1 lists the resultant estimates of near-wall conductivities, $k_{m,w}$, in terms of
its ratio to the conductivity of water. No significant difference from the conductivity
of water is observed unless the solid phase material has a very different value, the
steel-water case here.

It can be concluded that the near-wall conductivities are in fact dominated by the
value for the fluid phase material. Even if experimental values for randomly packed
beads were used, the values would be arithmetically closer to those of the fluid than
to those of the solid. This trend is more significant near an impermeable wall. The
present study therefore does not confirm that the material of the solid phase has an
effect on near-wall conductivities by comparisons of the temperature profiles from
the experiments and conjugate numerical simulations.

For the estimate of thermal dispersion, error bounds for the stagnant conductivities must be established. The lower bound is the value of the fluid phase, and the
upper bound is that based on the experimental observations assuming random

packing. The medium comprising polyethylene spheres and water is an exception because polyethylene has a lower conductivity than water. Their values are close, so the lower bound is the value for polyethylene and the upper value is that of water.

6.2 Momentum Transport

The dynamics of momentum transport is closely coupled to the transport of energy in buoyancy-driven convection. The momentum equation for porous media (beginning with Darcy's experiments) has grown to a semi-heuristic model which connects Darcy's model to Navier-Stokes equations.

Difficulties include the effects of the impermeable wall, which is typically represented by the Brinkman term with a modified near-wall viscosity. However, the near wall momentum transport problem does not appear to be completely resolved in the literature. One reason for this situation is the validity of volume-averaging in this region. This is not an easy question to answer because it cannot be answered geometrically. The validity of volume averaging depends also on flow conditions, which depend on thermal conditions in a buoyancy-driven flow.

Our results confirm that the gravity-extended Darcy's flow model is suitable for most of the cases, based on steady-state measurements. This claim can be justified for a glass-water porous medium. Glass and water have thermal conductivities that are close to each other, which minimizes uncertainties in the stagnant conductivity. The experiments using both 1.5 and 6 mm DIA glass beads result in a regression exponent of 0.4 for the Nusselt number as a function of the Rayleigh-Darcy number. This exponent can be obtained by a simple scale analysis and by numerical simulation [89, 90]. The possible effects of thermal dispersion need to be addressed simultaneously. However, they can be explained by the slight shift in the experimental data points from the region bounded by Bejan's correlation.

6.3 A Model for Thermal Dispersion

Bejan's analysis [89] for a two-dimensional enclosure results in a uniform boundary layer thickness and an exponential velocity and temperature profile in the boundary layers. The maximum velocity therefore is the same in the longitudinal direction. Based on the analysis developed in Chap. 3, thermal dispersion is most important where there is a large velocity gradient, and this is in the vicinity of the wall in the present study. The fact that Bejan's correlation predicts the exponent very well implies that his analysis has the correct set of terms in the governing equations. Actual values of the thermophysical properties shift the correlation accordingly, without changing the slope.

A significant component of the dispersion tensor, Eqn. (3.41), may be assumed to be a function of pore-scale Peclet number, Eqn. (3.42), and an unknown function, $\mathcal{A}$,

$$\frac{\alpha_{dis}}{\alpha_f} = \mathcal{A} \cdot Pe_p \tag{6.1}$$

The pore-scale Peclet number, Pe_p, is expected to explain the fluctuating velocity component that drives thermal dispersion by its relation to the intrinsic volume-averaged velocity, $\left\langle \vec{u} \right\rangle^f$, for a homogeneous periodic porous medium. The actual medium that is being investigated here does not strictly meet these assumptions. However it is possible, if the bead diameter is small enough so that the velocity boundary layer along a flat surface fully interacts with the tortuous flow paths in the neighboring porous medium, the boundary layer becomes three-dimensional enough, and the pore-scale Peclet number well represents the conditions. In other words, these effects are to be explained by the unknown function, $\mathcal{A}$, based on measurement.

Although the porous medium is not homogeneous near the wall, the geometry can be expected to be periodic in the longitudinal direction, i.e., along the wall. There-fore, the unknown function $\mathcal{A}$ can be expected to be constant. With Bejan's [89] result of uniform velocity, the dispersivity, Eqn. (6.1), becomes a constant including the Peclet number. Constant thermal dispersivity allows one to combine the diffu-sion and the dispersion terms in the energy equation by an arithmetic addition of thermal diffusivity and the dispersivity. This serves as an increase in the apparent diffusivity for the overall problem of the present investigation. The resulting Nusselt number is then expected to be shifted to the right from an analytical solution because the Rayleigh-Darcy number uses only the diffusivity of the fluid, which is less than the sum of the diffusivity and the dispersivity. The magnitude of the dispersivity, in other words, can be deduced by adjusting the Rayleigh-Darcy numbers for the experimental data so that the data points would line up with the correlation obtained from analysis.

An appropriate velocity scale is necessary when calculating a pore-scale Peclet number. Based on the velocity profile in [89], the velocity at the wall can be calculated by,

$$u_w = \frac{\alpha_m}{H} \left(\frac{H}{L}\right)^{-0.2} (Ra^*Da)_H^{0.6} \tag{6.2}$$

where H is the height of the recirculation domain, as well as that of heated side wall, and L is the depth of the domain. For the present investigation, H = 350 mm, and H/L ≈ 3.5.

Figure 6.2 shows the distribution of the total dispersion coefficient, α_{total}, where

$$\alpha_{total} = \alpha_m + \alpha_{dis} \tag{6.3}$$

The fluid is water for all the cases. For each bead type, there is a general trend of increasing total dispersion coefficient with Peclet number, except for the 1.5 mm diameter glass beads. The literature on total dispersion coefficient as a function of Peclet number for randomly packed porous media generally confirms that the ratio of

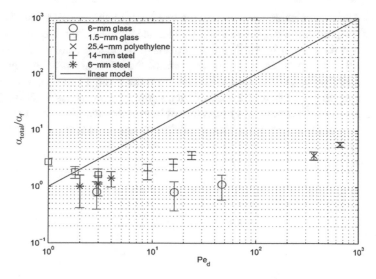

Fig. 6.2 Total dispersion coefficient, $\alpha_{total} = \alpha_m + \alpha_d$. The velocity scale for Peclet number is based that of [89]. Water is the fluid

total dispersion coefficient to fluid diffusivity remains <1 for Pe < 1 and increases for Pe > 1, and similar trends are observed here. If the results for randomly packed media are to be directly applied for comparison, however, the magnitudes of the total dispersion coefficient for 14 and 25.4 mm DIA polyethylene cases are low for the Peclet numbers. This may be because, Eqn. (6.2), does not apply. As discussed in Chap. 5, these large-diameter cases have Nusselt numbers that are better explained by a correlation for fluid convection. Large bead diameter causes the estimates of permeability and pore-velocity to be high.

Figure 6.2 also shows a general difference between the results for the 6 mm DIA glass and steel beads. As expected, the 6 mm DIA steel case has a larger total dispersion coefficient, which is caused by a thermal conductivity that is larger than that of glass. The difference in Peclet numbers is noticeable as well. The steel beads show the effects of thermal dispersion at relatively low Peclet number.

The effect of bead diameter can be addressed by comparing the 1.5 mm DIA and 6 mm DIA glass cases. The reason that the 1.5 mm DIA beads produce a larger dispersion coefficients is perhaps due to boundary layer interaction with more than the first layer of beads. The 6 mm DIA case not only has much fewer contact points with the vertical wall but also may provide a larger fluid volume for boundary-layer type flow to form. The opposite trend is observed for the 6 and 14 mm DIA steel cases. The reason is not clear, but it may be related to the application of porous medium analysis to the 14 mm DIA case.

The data presented in Fig. 6.2 show an increase in dispersion with Peclet number, but the magnitudes, especially at large Peclet number, are small compared to those of the linear model, which in the literature appear based on a homogeneous porous media. The basis for the small magnitudes is most likely the high local porosity near the wall. The impermeable flat wall and the first layer of spheres all aligned along it create a flow path that is much less tortuous than one far away from the wall. The fact that the large diameter cases show greater effects of dispersion may be related to the size of eddies that result from the interruption of the boundary layer by the beads. This trend is observed in the comparison between the 6 and 14 mm DIA steel cases and the comparison between the polyethylene and glass bead cases. Polyethylene has a smaller thermal conductivity than that of both water and glass, but their values are close so that the resulting stagnant conductivity is near to that of the glass-water system.

Chapter 7
Conclusion

Heat transfer in fluid-saturated porous media has been investigated with the objective to estimate the extent of thermal dispersion in the vicinity of a vertical impermeable wall. The volume-averaging method following Whitaker's [73] approach has been employed to analytically explain thermal dispersion. Experiments have been conducted to seek and observe any indication of thermal dispersion as shown in wall temperature variation.

Thermal dispersion can be qualitatively defined as a mixing of thermal energy due to microscopic flow through tortuous paths in a porous medium. This phenomenon has been mathematically defined through a derivation of the energy equation for a homogeneous porous medium under local thermal equilibrium. The derivation shows clearly that thermal dispersion is separate from the effective diffusive energy transport, or conduction, and could be significant, depending on the microscopic velocity and temperature fields. The volume-averaging technique allows properties to be written in terms of their volume-averaged quantities. The technique eliminates equations written for microscopic variables and simplifies the problem in many respects. However, microscopic velocity and temperature remain in the volume-averaged energy equation and create a closure problem. Following Whitaker [73] on a closure for mass dispersion, thermal dispersion is modeled as a function of the gradient of volume-averaged temperature. Stagnant thermal conductivity, which contains a surface integral, is modeled in a similar way, which makes the final form of the energy equation, Eqn. (3.38), similar to that proposed by Cheng (Eqn. (2.23)) [45]. The present analysis results in a relatively simple energy equation but has necessitated that a new coefficient, $\overline{\overline{A}}_{\text{dis}}$, (Eqn. (3.41)) be determined experimentally.

Our experiments have achieved several objectives. One that is the most important is to estimate thermal dispersion in the region near an impermeable wall. This region is where an assumption of homogeneous porous medium may not be valid, and studies focused on this region are not very common in the literature. The second is to study buoyancy driven flow along a vertical wall which is not very common in the

H. Sakamoto, F. A. Kulacki, *Buoyancy-Driven Flow in Fluid-Saturated Porous Media near a Bounding Surface*, SpringerBriefs in Applied Sciences and Technology, https://doi.org/10.1007/978-3-319-89887-2_7

literature but is important in estimating the dispersion. In particular, Nusselt numbers were obtained as functions of Rayleigh-Darcy number for porous media composed of steel, glass and polyethylene, while water as the interstitial fluid. Thirdly, the technique have been used is to measure temperature non-invasively such that the flow field developed in the porous medium is entirely due to the presence of the selected solid phase material. Finally, the selection of solid phase materials allows a separation of dispersive effects from diffusion in the near wall region.

The solid phase of the porous media are made of spherical beads, which is a common approach seen in the literature. The materials chosen for the solid phase are borosilicate glass, steel and polyethylene. The glass has a thermal conductivity nearly the same as that of water, which eliminates the difficulty of estimating the stagnant conductivity. Chrome steel has a thermal conductivity that is approximately 100 times that of water. The stagnant conductivity was estimated from published data in the literature for a medium made of steel beads and water, and the high conductivity ratio allowed us to estimate how the solid phase participates in heat transfer. This is particularly interesting in the vicinity of an impermeable wall because of the high and variable porosity in that region. Polyethylene has a lower thermal conductivity than that of water. However, the stagnant conductivity of a porous medium made of polyethylene beads and water was not expected to be very different from that of a medium made of glass and water because of the logarithmic nature of the stagnant conductivity as a function of the conductivity ratio.

Bead diameters of the beads are 1.5 and 6.0 mm DIA for glass, 6.0 and 14.0 mm DIA for steel and 25.4 mm for polyethylene. The medium made of 6.0 mm DIA glass beads and 6.0 mm DIA steel beads allowed examination the effects of solid phase conductivity. Different bead diameters of the same solid phase material allowed comparison of the influence of the diameter with regard to its interaction with the boundary layer, which is approximately on the order of 1.0 mm for water alone. Large polyethylene beads can be seen as an extension of glass because of its similar conductivity.

An important finding from transient data is that temperature profiles of the wall in the conduction regime are not affected significantly by the estimate of stagnant conductivity of the medium. If stagnant conductivity is not known, the fluid value can be used, at least for glass-air, polyethylene-air and steel-air media while the solid phase is packed spherical beads.

Measurement of steady state heat transfer has resulted in the following key findings:

- Heat transfer correlations proposed by Cheng and Minkowycz [41] and Bejan [89] are validated. The present data suggest a slope that is closer to that suggested by Bejan [89]; however, experimental uncertainty prevents rejecting the Cheng-Minkowycz correlation.
- Weak effects of thermal dispersion are observed for the saturated porous media investigated, and the data lie close to correlations developed assuming no dispersive effects.

- The steady-state heat transfer correlation lies on a straight line in a log-log field, and nonlinear behavior such as the one observed by Imadojemu and Johnson [50] is not observed.
- The volume-averaging method is valid even in the vicinity of an impermeable wall for the media studied in the present study.

Appendix A
Volume Averaging Theorems

Following Hubbert's [3] attempt to explain the empirically measured velocity as a volume-averaged quantity, a theoretical approach to the problem clearly became necessary. Slattery [4] derives the volume-averaged momentum equation, and Whitaker [5] presents a more thorough derivation, clearly pointing out the mathematical theorem on which the volume-averaging method is based.

Consider an arbitrary function, ψ, which may be density, velocity or temperature, defined for the fluid phase. The volume-averaged quantity is defined as follows.

$$\langle \psi \rangle = \frac{1}{V_{REV}} \int_{V_f} \psi dV \qquad (A.1)$$

where V_{REV} is the representative elementary volume that is indicated by a circle in Fig. A.1. The integration is over the fluid phase since the arbitrary function is defined only for this phase. We can examine how the integrated quantity changes spatially, assuming that the size of REV remains constant.

If a neighboring REV is displaced a small distance Δs from its original location, an equation that is analogous to the Reynolds transport theorem can be written,

$$\frac{d}{ds} \int_{V_f} \psi dV = \int_{V_f} \left(\frac{\partial \psi}{\partial s} \right) dV + \int_{A_f} \psi \left(\frac{\partial \vec{r}}{\partial s} \right) \cdot \hat{n} dV, \qquad (A.2)$$

where A_f is the interface area of the fluid phase that faces either the solid phase or the boundary of the REV (i.e., $A_f = A_{sf} + A_i$, where A_i is the area at the boundary of REV that is not solid). Comparing this equation with Fig. A.1 at a given time, while focusing on the left most channel, the change in the integral of the function (l.h.s) is the sum of the function contained in the area overlapped by both REVs (first term on the r.h.s) and the function newly contained in or left out of the REV (the last term).

© The Author(s), under exclusive licence to Springer International Publishing AG,   89
part of Springer Nature 2018
H. Sakamoto, F. A. Kulacki, *Buoyancy-Driven Flow in Fluid-Saturated Porous Media near a Bounding Surface*, SpringerBriefs in Applied Sciences and Technology, https://doi.org/10.1007/978-3-319-89887-2

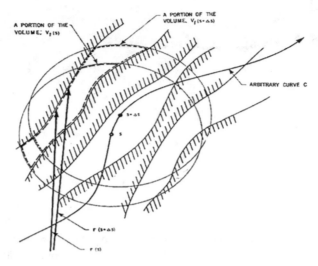

Fig. A.1 Schematic porous medium indicating representative elementary volume for the volume-averaging theorem [5]

At a given time, however, the first term on the right-hand side vanishes. It can be shown that the above equation then yields,

$$\nabla \int_{V_f} \psi dV = \int_{A_i} \psi \widehat{n} dA \tag{A.3}$$

where $\widehat{n}$ is the outward unit normal on area elements A_i. The volume integral of the gradient of the function can be written as follows, using the Green's theorem,

$$\int_{V_f} \nabla \psi dV = \int_{A_i} \psi \widehat{n} dA + \int_{A_{fa}} \psi \widehat{n} dA. \tag{A.4}$$

Substituting Eqn. (A.3), yields,

$$\int_{V_f} \nabla \psi dV = \nabla \int_{V_f} \psi dV + \int_{A_{fa}} \psi \widehat{n} dA \tag{A.5}$$

Dividing this equation by the volume of the REV and changing the notation to the volume average as shown in Eqn. (A.1), yields the volume averaging theorem for a gradient,

$$\langle \nabla \psi \rangle = \nabla \langle \psi \rangle + \frac{1}{V_{REV}} \int_{A_{fs}} \psi \widehat{n} dA. \tag{A.6}$$

where the volume of the REV is assumed constant.

A consequence of Eqn. (A.6) is that the volume averaging theorem for a divergence, which can be written,

$$\langle \nabla \cdot \psi \rangle = \nabla \cdot \langle \psi \rangle + \frac{1}{V_{REV}} \int_{A_{fs}} \psi \cdot \hat{n} dA. \tag{A.7}$$

Equations (A.6) and (A.7) clearly show that the gradient (or divergence) of an averaged quantity is not necessarily the same as the average of the gradient (or the divergence) of the quantity.

Appendix B
Uncertainty Analysis

The Nusselt number is given by Eqn. (4.8) in term of measured quantities, and the uncertainty is estimated by [91],

$$(\delta \mathrm{Nu}_x)^2 = \left(\frac{\partial \mathrm{Nu}_x}{\partial q_w''} \cdot \delta q_w''\right)^2 + \left(\frac{\partial \mathrm{Nu}_x}{\partial x} \cdot \delta x\right)^2 + \left(\frac{\partial \mathrm{Nu}_x}{\partial k_m} \cdot \delta k_m\right)^2$$
$$+ \left(\frac{\partial \mathrm{Nu}_x}{\partial \Delta T} \cdot \delta(\Delta T)\right)^2 \tag{B.1}$$

where $\Delta T = T_w(x) - T_\infty$. Similarly, the uncertainly in the flux-based Rayleigh-Darcy number is,

$$\left[\delta (\mathrm{Ra}_m \mathrm{Da})_x^*\right]^2 = \left[\frac{\partial (\mathrm{Ra}_m \mathrm{Da})_x^*}{\partial \beta} \cdot \delta \beta\right]^2 + \left[\frac{\partial (\mathrm{Ra}_m \mathrm{Da})_x^*}{\partial q_w''} \cdot \delta q_w''\right]^2$$
$$+ \left[\frac{\partial (\mathrm{Ra}_m \mathrm{Da})_x^*}{\partial x} \cdot \delta x\right]^2 + \left[\frac{\partial (\mathrm{Ra}_m \mathrm{Da})_x^*}{\partial K} \cdot \delta K\right]^2 \tag{B.2}$$
$$+ \left[\frac{\partial (\mathrm{Ra}_m \mathrm{Da})_x^*}{\partial \nu} \cdot \delta \nu\right]^2 + \left[\frac{\partial (\mathrm{Ra}_m \mathrm{Da})_x^*}{\partial \alpha_m} \cdot \delta \alpha_m\right]^2 \left[\frac{\partial (\mathrm{Ra}_m \mathrm{Da})_x^*}{\partial k_m} \cdot \delta k_m\right]^2$$

The uncertainty of the underlying quantities in the Eqns. (B.1) and (B.2) can be written in terms of percentage errors. For example, for the wall heat flux for Nusselt number, the partial derivative is,

$$\frac{\partial \mathrm{Nu}_x}{\partial q_w''} = \frac{x}{k_m[T_w(x) - T_\infty]}. \tag{B.3}$$

The first term on the r.h.s. of Eqn. (B.1) becomes,

© The Author(s), under exclusive licence to Springer International Publishing AG, part of Springer Nature 2018
H. Sakamoto, F. A. Kulacki, *Buoyancy-Driven Flow in Fluid-Saturated Porous Media near a Bounding Surface*, SpringerBriefs in Applied Sciences and Technology, https://doi.org/10.1007/978-3-319-89887-2

$$\left(\frac{\partial \text{Nu}_x}{\partial q_w''} \cdot \delta q_w''\right)^2 = \left(\frac{q_w'' x}{k_m[T_w(x)-T_\infty]} \frac{\delta q_w''}{q_w''}\right)^2.$$
$$= \text{Nu}_x^2 \left(\frac{\delta q_w''}{q_w''}\right)^2 \tag{B.4}$$

Similarly, the term for stagnant conductivity yields,

$$\left(\frac{\partial \text{Nu}_x}{\partial k_m} \cdot \delta k_m\right)^2 = \left(\frac{-q_w'' x}{k_m^2[T_w(x)-T_\infty]} \cdot \delta k_m\right)^2$$
$$= \left(\frac{q_w'' x}{k_m[T_w(x)-T_\infty]} \cdot \frac{-\delta k_m}{k_m}\right)^2. \tag{B.5}$$
$$= \text{Nu}_x^2 \left(\frac{\delta k_m}{k_m}\right)^2$$

Substituting these results into Eqn. (B.1),

$$\left(\frac{\delta \text{Nu}_x}{\text{Nu}_x}\right)^2 = \left(\frac{\delta q_w''}{q_w''}\right)^2 + \left(\frac{\delta x}{x}\right)^2 + \left(\frac{\delta k_m}{k_m}\right)^2 + \left(\frac{\delta \Delta T}{\Delta T}\right)^2. \tag{B.6}$$

For the Rayleigh-Darcy number, the non-linear term due to longitudinal x-location can be written,

$$\left[\frac{\partial (\text{Ra}_m \text{Da})_x^*}{\partial x} \cdot \delta x\right]^2 = \left[2 \frac{g\beta q_w'' x K}{\nu \alpha_m k_m} \cdot \delta x\right]^2$$
$$= \left[\frac{g\beta q_w'' x^2 K}{\nu \alpha_m k_m} \cdot 2\frac{\delta x}{x}\right]^2 \tag{B.7}$$
$$= \left[(\text{Ra}_m \text{Da})_x^*\right]^2 \left(2\frac{\delta x}{x}\right)^2$$

The resulting expression for the percentage error in Rayleigh-Darcy number is,

$$\left[\frac{\delta (\text{Ra}_m \text{Da})_x^*}{(\text{Ra}_m \text{Da})_x^*}\right]^2 = \left(\frac{\delta \beta}{\beta}\right)^2 + \left(\frac{\delta q_w''}{q_w''}\right)^2 + \left(2\frac{\delta x}{x}\right)^2 + \left(\frac{\delta K}{K}\right)^2$$
$$+ \left(\frac{\delta \nu}{\nu}\right)^2 + \left(\frac{\delta \alpha_m}{\alpha_m}\right)^2 + \left(\frac{\delta k_m}{k_m}\right)^2. \tag{B.8}$$

Uncertainty in Longitudinal x-Locations

The sources of this uncertainty include the finite diameter (1.6 mm) of thermocouple embedding holes, the location of the centerline of each hole, and the true location of the leading edge. Of these, the uncertainty in the location of the leading edge may be the dominant factor. Also, these uncertainties are in physical dimensions, so greater percentage errors result for locations close to the leading edge.

The holes that hold thermocouples are made using a 0.0625 in. (~1.6 mm DIA) drill. A thermocouple bead can be located anywhere in the hole, which gives an uncertainty of ±0.8 mm. When machining the holes, the center location of each is

Table B.1 Uncertainty in longitudinal location

x (m)	δx (m)	$\delta x/x$
0.0015	0.004	2.62
0.0254	0.004	0.16
0.0508	0.004	0.08
0.0762	0.004	0.05
0.1016	0.004	0.04
0.1270	0.004	0.03
0.1524	0.004	0.03
0.1778	0.004	0.02
0.2032	0.004	0.02

measured from the leading edge of the brass plate. This produces an uncertainty of approximately ±1 mm. When assembling the plates, however, the leading edge is glued with silicone sealant with a thickness of approximately 3 mm. Assuming that the true leading edge is located between the edge of the sealant and the edge of the brass plate, the uncertainty is +3/−0 mm. Assuming that a combined effect of these three uncertainty results in +4 mm, Table B.1 summarizes estimated uncertainty in longitudinal location, x, for each of the thermocouple locations.

Uncertainty in Thermal Conductivities

The stagnant conductivity of the porous medium and water are important properties for this investigation. Uncertainty in the stagnant thermal conductivity is estimated based on the experimental investigations summarized in Fig. 4.9. That of water is estimated based on the variation over a range of temperature.

The uncertainty in the conductivity of water should be discussed first because stagnant conductivities are normalized by the fluid conductivity and Eqn. (4.5). Figure B.1 shows the thermal conductivity of water at several temperatures [92]. The solid line fits the data in the range showing an approximately linear profile. In the present investigation, we assume constant properties evaluated at the average film temperature. The maximum error in the conductivity of water at 300 K is ~10%, as indicated in the figure and assuming that the average temperature difference between the plate and the far field is ~50 K. This occurs in the case of high heat flux with 1-mm glass beads [77]. However, for almost all other experiments, the temperature difference is limited to ~20 K for which the error is less than 5%.

Stagnant conductivities are evaluated using Fig. 4.9 and Eqn. (4.5). The normalizing conductivity of water is evaluated at 300 K, and we assume that the data presented in Fig. 4.6 are exact at 300 K because the temperature-dependency of the stagnant conductivity is expected to be weak. Taking into account both the spread in the experimental data and the uncertainty of 5% in the conductivity of water, we estimate the uncertainties for the conductivity of the porous medium, k_m, as shown in Table B.2. The large uncertainty of the steel-water porous medium results from the

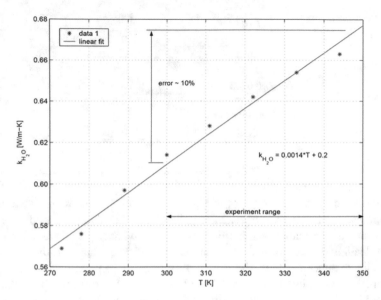

Fig. B.1 Variation of water conductivity with temperature [92] and estimate of error due to the use of the conductivity at 300 K

Table B.2 Uncertainty in stagnant thermal conductivity

	k	$\delta k_m/k_m$
Water	–	0.05
Glass-water	1.0	0.05
Steel-water	100	0.40
Polyethylene-water	0.54	0.05

spread of experimental data at that high conductivity ratio, $\kappa = 100$. It should be noted that the estimated uncertainty is for a relatively homogeneous medium, not for a region near an impermeable wall.

Uncertainty in Temperature

The sources of uncertainty in measured temperature include that in the reference thermometer used in calibration, the linear correlation for calibration data, and the small voltage measurement and fluctuation for each thermocouple. Temperature measurements were occasionally consistently different between thermocouples before experiments are run. This was observed more frequently after experiments are run multiple times and may be related to a changing property of the embedded thermocouples themselves. However, the differences are consistent throughout an experiment, and it is possible to correct them based on the differences before the data

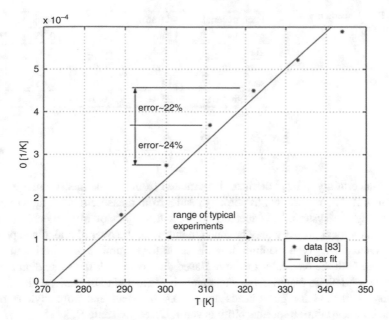

Fig. B.2 Variation of coefficient of volumetric expansion of water [93] and estimate of error for typical experiments

run. These differences are observed to be ±0.5 °C and are a dominating factor for the uncertainty in measured temperature differences, ΔT.

Uncertainty in the Coefficient of Volumetric Expansion

Figure B.2 shows temperature variation of the volumetric expansion coefficient, β. An uncertainty of 24% is assumed for an average temperature difference ΔT = 20 K. Because the expansion coefficient varies approximately linearly with temperature, the uncertainty is assumed to vary linearly with temperature. For example, an uncertainty of 12% is assumed for an average temperature difference ΔT = 10 K.

Uncertainty in Porosity and Permeability

Permeability is calculated using Eqn. (4.2), and total uncertainty results from that in porosity and the bead diameter,

Table B.3 Uncertainty in permeability

Bead material	d (mm)	δd/d	δφ/φ	δK/K
glass	1.5	0.15	0.05	0.37
glass	6.0	0.02	0.05	0.23
steel	6.0	0.01	0.05	0.22
steel	14.0	0.01	0.05	0.22
polyethylene	25.4	0.01	0.05	0.22

$$\left(\frac{\delta K}{K}\right)^2 = \left(2\frac{\delta d}{d}\right)^2 + \left(\frac{3-\phi}{1-\phi}\frac{\delta\phi}{\phi}\right)^2. \tag{B.9}$$

The uncertainty in bead diameter is estimated from sample measurements shown in Table 4.2. The uncertainty in porosity results from measurement of the volume of fluid filling the system. This method neglects the effect of impermeable surfaces along which local porosity may be much higher than the overall one. Once porosity is measured for a given medium, it is the same for a number of experimental runs conducted with the particular medium. Based on media with the same diameter and the same solid phase material, the uncertainty is estimated to be ±2%, while the bulk porosity is ≈ 0.38 for glass beads and ≈ 0.4 for steel and polyethylene beads. Estimated uncertainty in permeability is summarized in Table B.3.

Uncertainty in Water Properties

The kinematic viscosity of water and thermal diffusivity of the porous medium based on volumetric heat capacity of water are dependent on temperature. Based on an overall temperature difference of 20 K, the variation is as much as 25%. Because kinematic viscosity is approximately a linear function of temperature in the small temperature range of the present investigation, the error can be assumed to vary in the same manner. For example, an uncertainty of 12.5% is estimated for a temperature difference of 10 K.

The contribution of volumetric heat capacity to the uncertainty of thermal diffusivity is limited to <0.5% if an overall temperature difference of 20 K is assumed. Therefore, the uncertainty of stagnant conductivity dominates that of the porous medium thermal diffusivity (Table B.2). Again, temperature dependency of the stagnant conductivity is assumed to be negligible.

Overall Uncertainty in $\left(\mathrm{Ra_m^*Da}\right)_x$

Substituting the individual uncertainties into Eqns. (B.1) and (B.3), Table B.4 results. For a given heat flux, only two longitudinal x-locations are presented to show the effects of the different uncertainties in x.

Table B.4 Overall uncertainties

Solid (d)	$q_w''\left(\text{W}/\text{m}^2\right)$	x (m)	$\delta x/x$	$\delta\Delta T/\Delta T$	$\delta Nu/Nu$	$\dfrac{\left(\text{Ra}_m^*\text{Da}\right)_x}{\left(\text{Ra}_{\text{total}}^*\text{Da}\right)_x}$
Glass (6.0 mm)	1240	0.0254	0.16	0.067	0.18	0.42
	1240	0.0762	0.05	(7.5)	0.10	0.29
	5160	0.0254	0.16	0.028	0.17	0.51
	5160	0.0762	0.05	(17.8)	0.08	0.40
	80	0.0254	0.16	0.345	0.38	0.40
	80	0.0762	0.05	(1.45)	0.35	0.26
Glass (1.5 mm)	5360	0.0254	0.16	0.010	0.17	1.02
	5360	0.0762	0.05	(51.5)	0.08	0.97
	1270	0.0254	0.16	0.017	0.17	0.72
	1270	0.0762	0.05	(30.1)	0.08	0.65
Steel (14.0 mm)	1260	0.0254	0.16	0.127	0.45	0.69
	1260	0.0762	0.05	(4.0)	0.42	0.62
	5190	0.0254	0.16	0.049	0.43	0.71
	5190	0.0762	0.05	(10.3)	0.41	0.64
	2820	0.0254	0.16	0.077	0.44	0.70
	2820	0.0762	0.05	(6.5)	0.41	0.63
Steel (6.0 mm)	1240	0.0254	0.16	0.096	0.44	0.69
	1240	0.0762	0.05	(5.2)	0.42	0.62
	2900	0.0254	0.16	0.049	0.43	0.71
	2900	0.0762	0.05	(10.3)	0.41	0.64
	5160	0.0254	0.16	0.035	0.43	0.73
	5160	0.0762	0.05	(14.4)	0.41	0.66
Polyethylene (25.4 mm)	1230	0.0254	0.16	0.111	0.20	0.40
	1230	0.0762	0.05	(4.5)	0.13	0.27
	2990	0.0254	0.16	0.060	0.18	0.42
	2990	0.0762	0.05	(8.4)	0.10	0.29

$\delta_w''/q_w'' = 0.03$ is taken from Fig. 4.8, and $\delta k_m/k_m$ is based on Table B.2. ΔT is based on an average of the highest and the lowest temperatures [77]. $\delta K/K$ is based on Table B.3. The fluid is water. See [77] for additional detail

Uncertainty in Dispersion Coefficient

The total dispersion coefficient α_{total} (Fig. 6.2) is estimated comparing Rayleigh-Darcy numbers, $\left(\text{Ra}_m^*\text{Da}\right)_x$, between the present results and those of Bejan [89]. The ratio of total dispersion coefficient, to thermal diffusivity of fluid can be rewritten as,

$$\frac{\alpha_{\text{total}}}{\alpha_m} = \frac{\left(\text{Ra}_m^*\text{Da}\right)_x}{\left(\text{Ra}_{\text{total}}^*\text{Da}\right)_x}\frac{\alpha_m}{\alpha_f} \tag{B.10}$$

where $\text{Ra}_{x,\text{total}}^*$ is based on the total dispersion coefficient.

　　While obtaining the corresponding value for the Rayleigh-Darcy number based on total dispersion coefficient from Eqn. (4.11) for Nusselt number obtained in the present measurements, the uncertainty in the total diffusivity can be estimated by,

$$\left[\frac{\delta\left(\alpha_{\text{total}}/\alpha_f\right)}{\alpha_{\text{total}}/\alpha_f}\right]^2 = \left[\frac{\delta\left(\text{Ra}_m^*\text{Da}\right)_x}{\left(\text{Ra}_m^*\text{Da}\right)_x}\right]^2 + \left[\frac{\delta\alpha_m}{\alpha_m}\right]^2 + \left[\frac{\delta\alpha_f}{\alpha_f}\right]^2 \tag{B.11}$$

　　The uncertainties in Rayleigh-Darcy numbers are from Table B.4. The uncertainties in porous medium diffusivities and that of water are from Table B.2. Resulting uncertainties are shown in Fig. 6.2, along with the data.

References

1. Lapwood ER (1948) Convection of a fluid in a porous medium. Proc Camb Philos Soc 44:508–521
2. Nield DA, Bejan A (1992) Convection in porous media. Springer-Verlag, New York
3. Hubbert MK (1956) Darcy's law and the field equations of the flow of underground fluids. Trans Am Inst Mining Metal Pet Eng Pet Branch 207:222–239
4. Slattery JC (1967) Flow of viscoelastic fluids through porous media. AIChE J 13(6):1066–1071
5. Whitaker S (1969) Advances in theory of fluid motion in porous media. Ind Eng Chem 61 (12):14–28
6. Wooding RA (1957) Steady state free thermal convection of liquid in a saturated permeable medium. J Fluid Mech 2:273–285
7. Aichlmayr HT (1999) The effective thermal conductivity of saturated porous media. Master's thesis, University of Minnesota
8. Darcy H (1856) Determination des lois d'ecoulement de l'eau a travers le sable, Les Font. Pub de la Ville de Dijon, 590–594
9. Prasad V, Kulacki FA, Keyhani M (1985) Natural convection in porous media. J Fluid Mech 150:89–119
10. Nozad I, Carbonell RG, Whitaker S (1985) Heat conduction in multiphase systems. Ind Chem Eng Sci 40(5):843–855
11. Prasad V, Kladias N, Bandyopadhaya A, Tian Q (1989) Evaluation of correlations for stagnant thermal conductivity of liquid-saturated porous beds of spheres. Int J Heat Mass Transfer 32 (9):1793–1796
12. Waddams L (1944) The flow of heat through granular material. J Soc Chem Ind 63:337–340
13. Krupiczka B (1967) Analysis of thermal conductivity in granular materials. Int Chem Eng 7 (1):122–144
14. Jaguaribe EF, Beasley DE (1984) Modeling of the effective thermal conductivity and diffusivity of a packed bed with stagnant fluid. Int J Heat Mass Transfer 27(3):399–407
15. Preston FW (1957) (Unknown title). Ph.D. thesis, Pennsylvania State University
16. Lindfors J (1999) Boundary layer effects on the stagnant effective thermal conductivity of a saturated porous medium. Honors thesis in mechanical engineering, University of Minnesota
17. Bear J (1972) Dynamics of fluids in porous media. American Elsevier Publishing Co., New York
18. Kaviany M (1995) Principles of heat transfer in porous media, 2nd edn. Springer-Verlag, New York

© The Author(s), under exclusive licence to Springer International Publishing AG, 101
part of Springer Nature 2018
H. Sakamoto, F. A. Kulacki, *Buoyancy-Driven Flow in Fluid-Saturated Porous Media near a Bounding Surface*, SpringerBriefs in Applied Sciences and Technology, https://doi.org/10.1007/978-3-319-89887-2

19. Schneider KJ (1963) Investigation of the influence of free thermal convection on heat transfer through granular material. Proc. 11th Int. Cong. Refrigeration, H-4, pp 247–254
20. Elder JW (1967) Steady free convection in a porous medium heated from below. J Fluid Mech 27(1):29–48
21. Elder JW (1965) Chapter 8: Physical processes in geothermal areas. In: Lee WHK (ed) Geophysical monograph series: terrestrial heat flow, vol 8. Amer. Geophys. Union, Washington, pp 211–239
22. Kaneko T, Mohtadi MF, Aziz K (1974) An experimental study of natural convection in inclined porous media. Int J Heat Mass Transfer 17:485–496
23. Combarnous MA, Bia P (1971) Combined free and forced convection in porous media. Soc Petrol Eng J 11:399–405
24. Combarnous MA, Bories S (1974) Modelisation de la convection naturelle au sein d'une couche poreuse horizontale a l'laide d'un coefficient de transfert solide fluide. Int J Heat Mass Transfer 17:505–515
25. Cheng P (1985) Geothermal heat transfer. In: Rohsenow WM, Hartnett JP (eds) Handbook of heat transfer applications, 2nd edn. McGraw-Hill, New York, pp 11.1–11.54
26. Buretta RJ, Berman AS (1976) Convective heat transfer in a liquid saturated porous layer. J Appl Mech 98:249–253
27. Gupta BP, Joseph DD (1973) Bounds for heat transport in a porous layer. J Fluid Mech 57 (3):491–514
28. Catton I (1985) Natural convection heat transfer in porous media. In: Kakac A, Aung W, Viskanta R (eds) Natural convection: fundamentals and applications. Hemisphere, Washington, pp 514–547
29. Jakob M (1949) Heat transfer, vol 1. Wiley, New York
30. Yagi S, Kunii D (1957) Studies on effective thermal conductivities in packed beds. AIChE J 3 (3):373–381
31. Aichlmayr HT, Kulacki FA (2006) The effective thermal conductivity of saturated porous media. In: Green G et al (eds) Advances in heat transfer, vol 39. Academic, New York, pp 377–460
32. Batchelor GK, O'Brien RW (1977) Thermal or electrical conduction through a granular material. Proc R Soc Lond A 355:313–333
33. Yen Y-C (1974) Effects of density inversion on free convective heat transfer in porous layer heated from below. Int J Heat Mass Transfer 17:1349–1356
34. Hubbert MK (1940) The theory of ground-water motion. J Geol 48(8):785–944
35. Forchheimer P (1901) Wasserbewegung durch boden. Zeit Vereines Deut Ing 45 (1736–1741):1781–1788
36. Dupuit J (1863) Etudes theoriques et pratiques sur le mouvement des eaux dans les canaux decouverts et a travers les terrains permeabls, avec des considerations relatives au regime des grandes eaux, au debouche a leur donner, et a la marche des alluvions dans les rivieres a fond mobile, 2nd edn. Dunod, Paris
37. Brinkman HC (1949) A calculation of the viscous force exerted by a flowing fluid on a dense swarm of particles. Appl Sci Res A1:27–34
38. Beck L (1972) Convection in a box of porous material saturated with fluid. Phys Fluids 15 (8):1377–1383
39. Vafai K, Tien CL (1981) Boundary and inertia effects on convective mass transfer in porous media. Int J Heat Mass Transfer 25(8):1183–1190
40. Lage JL (1993) Natural convection within a porous medium cavity: predicting tools for flow regime and heat transfer. Int Comm Heat Mass Transfer 20(4):501–513
41. Cheng P, Minkowycz WJ (1977) Free convection about a vertical flat plate embedded in a porous medium with application to heat transfer from a dike. J Geophys Res 82(14):2040–2044
42. Wooding RA (1963) Convection in a saturated porous medium at large Rayleigh number or Peclet number. J Fluid Mech 15:527–544

43. Oberbeck A (1879) Ueber die warmeleitung der flussigkeiten bei berucksichtigung der stromungen infolge von temperaturdifferenzen. Ann Phys Chem 7:271–292
44. Boussinesq J (1901) Theorie analytique de la chaleur: mise en harmonie avec la thermodynamique et avec la theorie mecanique de la lumiere. Gauthier-Villars, Paris
45. Cheng P (1985) Natural convection in a porous medium: external flows. In: Kakac A, Aung W, Viskanta R (eds) Natural convection: fundamentals and applications. Hemisphere, Washington, pp 475–513
46. Masuoka T, Yokote Y, Katsuhara T (1981) Heat transfer by natural convection in a vertical porous layer. Bul JSME 24(192):995–1001
47. Jannot M, Naudin P, Viannay S (1973) Convection mixte en milieu poreux. Int J Heat Mass Transfer 16(2):395–400
48. Masuoka T (1968) A study of the free convection heat transfer about a vertical flat plate embedded in a porous medium. Trans JSME 34(259):491
49. Ofuchi K, Kunii D (1965) Heat-transfer characteristics of packed beds with stagnant fluids. Int J Heat Mass Transfer 8(5):749–757
50. Imadojemu H, Johnson R (1991) Convective heat transfer from a heated vertical plate surrounded by a saturated porous medium. Proc, ASME/JSME thermal eng. joint conf. pp 203–212
51. Bejan A, Poulikakos D (1984) The non-Darcy regime for vertical boundary layer natural convection in a porous medium. Int J Heat Mass Transfer 27(5):717–722
52. Huenefeld JS, Plumb OA (1981) Study of non-Darcy natural convection from a vertical heated surface in a saturated porous medium. Trans ASME, 81-HT-45
53. Ingham DB, Brown SN (1986) Flow past a suddenly heated vertical plate in a porous medium. Proc R Soc Lond A 403:51–80
54. Haq S, Mulligan JC (1990) Transient free convection about a vertical flat plate embedded in a saturated porous medium. Numer Heat Transfer A 18:227–242
55. Rees DAS, Pop I (2000) Vertical free convective boundary-layer flow in a porous medium using a thermal nonequilibrium model. J Porous Media 3(1):31–44
56. Taylor GI (1921) Diffusion by continuous movements. Proc Lond Math Soc 20:196–211
57. Reynolds O (1883) An experimental investigation of the circumstances which determine whether the motion of water shall be direct or sinuous, and of the law of resistance in parallel channels. Philos Trans R Soc Lond A 174:935–982
58. Koch DL, Brady JF (1985) Dispersion in fixed beds. J Fluid Mech 154:399–427
59. Gunn DJ, Khalid M (1975) Thermal dispersion and wall heat transfer in packed beds. Chem Eng Sci 30:261–267
60. Taylor GI (1953) Dispersion of soluble matter in solvent flowing slowly through a tube. Proc R Soc Lond A 219(1137):186–203
61. Aris R (1956) On the dispersion of a solute in a fluid flowing through a tube. Proc R Soc Lond A 235(1200):67–77
62. Yagi S, Kunii D, Wakao N (1960) Studies on axial effective thermal conductivities in packed beds. AICHE J 6(4):543–546
63. Cheng P (1981) Thermal dispersion effects in non-darcian convective flows in a saturated porous medium. Lett Heat Mass Transfer 8:267–270
64. Plumb OA (1983) The effect of thermal dispersion on heat transfer in packed bed boundary layers. Proc ASME-JSME Thermal Eng Conf 2:17–22
65. Jiang P-X, Ren Z-P, Wang B-X (1999) Numerical simulation of forced convection heat transfer in porous plate channels using thermal equilibrium and nonthermal equilibrium models. Numer Heat Trans A 35(1):99–113
66. Jiang P-X, Wang Z, Ren Z-P, Wang B-X (1999) Experimental research of fluid flow and convection heat transfer in plate channels filled with glass or metallic particles. Exp Thermal Fluid Sci 20(1):45–54
67. Kuo S, Tien CL (1988) Transverse dispersion in packed-sphere beds. Proc.1988 ASME-AIChE National Heat Trans. Conf. Vol 96:629–634

68. Wang B-X, Du J-H (1993) Forced convective heat transfer in a vertical annulus filled with porous media. Int J Heat Mass Transfer 36(17):4207–4421
69. Deleglise M, Simacek P, Binetruy C, Advani S (2003) Determination of the thermal dispersion coefficient during radial filling of a porous-medium. J Heat Transf 125(5):875–880
70. Kuwahara F, Nakayama A, Koyama H (1996) A numerical study of thermal dispersion in porous media. Trans ASME J Heat Transfer 118(3):756–761
71. Kuwahara F, Nakayama A (1999) Numerical determination of thermal dispersion coefficients using a periodic porous structure. Trans ASME J Heat Transfer 121(1):160–163
72. Whitaker S (1977) Simultaneous heat, mass, and momentum transfer in porous media: a theory of drying. Adv Heat Transf 13:119–203
73. Whitaker S (1999) The method of volume averaging. Kluwer Academic, Dordrecht, The Netherlands
74. Slattery JC (1972) Momentum, energy, and mass transfer in continua. McGraw-Hill, New York
75. Gray WG (1975) A derivation of the equations for multi-phase transport. Chem Eng Sci 30:229–233
76. Eidsath A, Carbonell RG, Whitaker S, Herrmann LR (1983) Dispersion in pulsed systems—III: comparison between theory and experiments for packed beds. Chem Eng Sci 38(11):1803–1816
77. Sakamoto H (2005) Buoyancy-driven flow in fluid-saturated porous media near a bounding surface. Doctoral dissertation, University of Minnesota, Minneapolis
78. Eckert ERG, Drake RM (1971) Analysis of heat and mass transfer. McGraw-Hill, New York
79. Bansal NP, Doremus RH (1986) Handbook of glass properties. Academic, Orlando
80. Lemmon E, McLinden M, Friend D (2003) Thermophysical properties of fluid systems: Water. In: Linstrom P, Mallard W (eds) NIST chemistry WEBBOok, NIST standard reference database number 69. National Institute of Standards and Technology, Gaithersburg. (http://webbook.nist.gov)
81. Kreith F, Bohn M (1993) Principles of heat transfer, 5th edn. West Publishing Company, St Paul, Co
82. Benenati R, Brosilow C (1962) Void fraction distribution in beds of sphere. AICHE J 8(3):359–361
83. Kristoffersen M (2001) Metastable convection in a porous medium heated from below. Master's thesis, University of Minnesota
84. Rees D, Pop I (2000) Vertical free convection in a porous medium with variable permeability effects. Int J Heat Mass Transfer 43(14):2565–2571
85. Ergun S (1952) Fluid flow through packed columns. Chem Eng Prog 48(2):89–94
86. Deissler RG, Boegli JS (1958) An investigation of effective thermal conductivities of powders in various gases. ASME Trans 80(7):1417–1425
87. Swift DL (1966) The thermal conductivity of spherical metal powders including the effect of oxide coating. Int J Heat Mass Transfer 9:1061–1074
88. Murthy PVSN, Singh P (1999) Heat and mass transfer by natural convection in a non-Darcy porous medium. Acta Mech 138(3-4):243–254
89. Bejan A (1983) The boundary layer regime in porous layer with uniform heat flux from the side. Int J Heat Mass Transfer 26:1339–1346
90. Degan G, Vasseur P (1996) Boundary-layer regime in a vertical porous layer with anisotropic permeability and boundary effects. Int. J. Heat Mass Transfer 18:334–343
91. Moffat RJ (1968), Planning experimental programs. Unpublished course notes. Department of Mechanical Engineering, Stanford University, Palo Alto
92. Electrical Research Association (1967) 1967 steam tables: thermodynamic properties of water and steam, viscosity of water and steam, thermal conductivity of water and steam. Edward Arnold Ltd., London

Printed in the United States
By Bookmasters